Green Energy and Technology

Climate change, environmental impact and the limited natural resources urge scientific research and novel technical solutions. The monograph series Green Energy and Technology serves as a publishing platform for scientific and technological approaches to "green"—i.e. environmentally friendly and sustainable—technologies. While a focus lies on energy and power supply, it also covers "green" solutions in industrial engineering and engineering design. Green Energy and Technology addresses researchers, advanced students, technical consultants as well as decision makers in industries and politics. Hence, the level of presentation spans from instructional to highly technical.

Indexed in Scopus.

Indexed in Ei Compendex.

John Lester Clarke

Sustainable Buildings for Sustainable Behaviour: Theory and Practice

Springer

John Lester Clarke
University of the Built Environment
Reading, Berkshire, UK

ISSN 1865-3529 ISSN 1865-3537 (electronic)
Green Energy and Technology
ISBN 978-3-031-98904-9 ISBN 978-3-031-98905-6 (eBook)
https://doi.org/10.1007/978-3-031-98905-6

This Springer imprint is published by the registered company Springer Nature Switzerland AG
The registered company address is: Gewerbestrasse 11, 6330 Cham, Switzerland

Acknowledgements This book is based on 25 years experience of study, volunteering, teaching and practice in the field of sustainable construction and environmental education culminating in Ph.D. research carried out through an Arts and Humanities Research Council funded Collaborative Doctoral Award between Kingston University, London and the Environment Trust for Richmond upon Thames.

I should firstly like to thank my Director of Studies, Dr. Stephen Pretlove for his invaluable guidance, expertise and professionalism throughout the research project. I would also like to thank all the staff and trustees at the Environment Trust for Richmond upon Thames for welcoming me into their organisation.

I also thank all those who participated in the case study surveys for their co-operation, patience, experience and wisdom, sharing their knowledge purely for altruistic reasons.

Finally, and by no means least, I would like to thank and dedicate this work to my parents for their constant love and support throughout.

As my beloved mother observed:

> A Ph.D. is a bit like doing the ironing; you've got to be in the mood!
>
> (Margery Clarke 2012)

Competing Interests The author has no competing interests to declare that are relevant to the content of this manuscript.

Contents

About the Author

John Lester Clarke graduated with an honour's degree in Building Surveying in 1992 and worked as a building surveyor for four years. In 1997 he was awarded a Post Graduate Certificate in Post–Compulsory Education (PGCE) and taught building technology. His interest in sustainability, education and the built environment grew along with wider environmental issues and in 1998 he gained a master's degree in Renewable Energy and Architecture from Nottingham University.

He has worked on a number of low environmental impact buildings in the UK, most notably at The Centre for Alternative Technology and the Far East including Japan and Thailand and has undertaken postgraduate and postdoctoral research and teaching across a wide spectrum of sustainability in the built environment at a number of UK universities. Working as an Environmental Education Project Officer he researched, developed and delivered education for sustainable development and was Project Manager for the design, construction, operation and use of an eco-building as a teaching and learning space for sustainable education.

John completed his Ph.D. in 2013, the inspiration and catalyst for this book. He then worked as a postdoctoral researcher in retrofitting buildings and developing low carbon technologies at University of Brighton. Since then, he has worked as deputy module leader and associate lecturer on undergraduate and postgraduate course related to sustainability in the built environment at University College of Estate Management.

His ongoing research and teaching interests are in sustainable architecture, natural construction materials, health and well-being in the built environment, education for sustainable development and pro-environmental behavioral change.

List of Figures

List of Tables

Chapter 1
Introduction

The broad aim of this book is to establish the extent to which sustainable buildings, throughout their design, construction, operation and use, can impact the long-term sustainable behaviour of individuals, organisations and institutions. This is explored through a number of factors which consider pedagogical, behavioural, technological and material interventions in the construction lifecycle in relation to social, environmental and economic sustainability.

The environmental impact of human behaviour on the design, construction and operation of buildings is often overlooked, frequently resulting in sub optimal performance over the lifecycle of the building (credibility gap and value-action gap). An over-reliance on technological and market solutions (false positivity) throughout the design, construction and operation of sustainable buildings means changing user behaviour is not currently envisaged by all but the most sustainability-minded built environment professionals. This study aims to develop an understanding of the dynamic and complex systems by which responsible environmental, social and economic action (sustainable behaviour) emerges from the relationship between people and the built environment.

The book title presents the topic to what extent sustainable buildings encourage sustainable behaviour, with broader topics covering the need for sustainable buildings and their social, environmental and economic benefits; a clear definition of sustainable behaviour and sustainable buildings; identifying opportunities for behavioural change from current best practice and how behavioural change theory can be applied to the built environment to encourage and optimise sustainable behaviour.

Existing literature on the topic reveals disparate levels of theory and practice in the fields of sustainability, architecture, behavioural psychology and pedagogy applied generally to the design, construction and operation of sustainable buildings. Five exemplar sustainable buildings with pedagogical and ecological functions are presented, highlighting common best practices and innovative approaches, as well as revealing barriers to achieving sustainable built environments that encourage sustainable behaviour.

J. L. Clarke, *Sustainable Buildings for Sustainable Behaviour: Theory and Practice*,
Green Energy and Technology, https://doi.org/10.1007/978-3-031-98905-6_1

The book reveals that there are numerous opportunities for behavioural interventions at critical stages throughout the lifecycle of buildings where 'value-action' gaps between our intentions to be more sustainable and our often sub-optimal actions or behaviours are identified. Strategies include education, information provision, training, experiential learning, feedback, participation and regulation.

The book contributes original knowledge by relating the way building mechanisms for change can be understood through the lens of behavioural psychology and the synthesis of the three disciplines of sustainability, architecture and pedagogy.

The book represents the study of the theory and practice of the development of low environmental impact buildings and the embedding of sustainable features, materials, technologies and processes in order to optimise their potential as resources for teaching and learning about sustainability and pro-environmental behavioural change. It is hoped this will not only lead to a more profound understanding of the impact of the built environment on the natural environment but will enable key stakeholders involved in the whole lifecycle of buildings throughout their procurement, design, construction and operation to positively change their behaviours in terms of environmental, social and economic sustainability.

A clear gap in knowledge exists around the problem of buildings, with varying levels of sustainability credentials, not achieving their optimum potential in terms of their technological, physiological and behavioural performance, identified as the 'credibility' and 'value-action' gaps resulting in sub-optimal social, financial and environmental outcomes.

Exemplar case studies are presented showing the development of five sustainable buildings; The WISE building, The DACE Eco-centre, Sidwell Friends School, The Core building and The Genesis Project. Each has environmental education and the promotion of sustainable building and lifestyle practices as key functions. An online survey, an analysis of project documentation and the assessment of buildings involving face-to-face interviews and questionnaires have been undertaken to support or refute the theoretical findings and highlight best practice and areas for improvement.

Orr [1], p. 128, a leading academic in this field, speaks of 'the hidden curriculum that is the building itself'. He states that despite great advancements in sustainable architecture in the past few decades, the majority of (unsustainable) existing and new buildings perform relatively poorly. By investigating current literature, it is shown that unsustainable buildings impart the message through their design, construction, operation and use that our consumption of scarce resources is disconnected from our everyday lives and they perpetuate the misconception throughout governments and societies that resources are somehow infinite, if they are considered at all.

It is generally the norm for us to be removed from natural processes in the buildings we inhabit such as the over reliance on artificial lighting, the increasing use of air conditioning and the increasing production and use of inorganic man-made materials. A sustainable building should be diametrically opposed to this and should in its design construction, operation and use both explicitly and implicitly encourage and enable us to lead more sustainable lifestyles.

Wastage, energy inefficiencies and unsustainable practices are often factored-in to building design and construction contracts. This is often reinforced through formal education provision, perpetuating unsustainable teaching into professional practice, a self-perpetuating closed loop of unsustainable practices.

A complex set of parameters exist that simultaneously both prevent and enable meaningful sustainable architecture to take place, i.e. architecture that achieves low environmental impact through the design, construction, operation and use of buildings allied with social and economic gains. These have been identified from the research as being pedagogical, technological, psychological, physiological, legislative and socio-economic. These parameters are investigated throughout the book from a theoretical perspective and from the practical application of sustainable design principles that considers sustainable human behaviour as a critical component in achieving a greater sustainable built environment.

There is now broad worldwide academic, scientific, social, political and economic consensus that the burning of fossil fuels impacts on our climate, affecting planetary temperatures with implications for flooding, extreme weather conditions, loss of biodiversity with associated negative economic consequences [2] if action is not taken urgently. Buildings play a significant part in this, accounting for up to 50% of total carbon emissions in the UK [3] through the extraction, manufacture, transportation and use of materials and energy expended in the construction and operation of buildings. This figure is around 40% globally but developing countries will soon reach or even exceed this figure if the low environmental impact design, construction, operation and use of buildings does not become standard practice world-wide.

Debates at the COP15 summit in Copenhagen in 2009 recognised the significance of buildings in tackling climate change. In a related article [4], p. 2 Dr Diana Urge-Vorsatz, director of the Center for Climate Change and Sustainable Energy Policy, stated 'widespread implementation of presently available technology and practices could reduce building-related emissions by between 40 and 70% by 2050.'

Climate change will bring new challenges for the global construction industry, and knowledge about potential impacts on the built environment will be of the utmost importance in years to come. The prospect of an even harsher climate means that we must pay more attention to the design, construction and location of buildings, and be more aware of the climatic impacts they will have to endure and adapt to. In order to ensure a resilient building stock in the future technical solutions have to be allied with greater sustainable behaviour with heightened awareness and realisation leading to action for sustainability requiring a significant shift in behavioural change.

An increased understanding of the need for sustainable buildings has driven the sustainable building agenda throughout the World. The need for energy efficient, low and even zero carbon buildings is widely recognised by governments and the built environment professions whilst building users are increasingly demanding more energy- efficient and sustainable buildings to live and work in, recognising social and environmental responsibilities as well as increasingly understood long term economic benefits.

However, there are strong social, political, economic and technological forces that prevent the adoption of sustainable construction practices and the behaviours of key stakeholders that continue to make the developed built environment unsustainable.

In a national context, the UK Government has previously set a number of challenging targets for improving sustainability [5], starting with the overarching goal of an 80% reduction in carbon emissions in the UK by 2050 (compared to 1990 levels) and recognising the need to change our building practices with targets for zero carbon and low water usage new-build housing and new build non-domestic buildings for England and Wales. Any discussion of sustainable buildings must include the need to tackle the existing building stock in terms of its environmental impact with energy efficient refurbishment and retrofitting requiring considerable behavioural change by individuals, organisations and wider institutions.

Clearly, in order to achieve these targets, we need to develop and utilise low carbon construction technologies and methods as well as fundamentally change the behaviour and attitudes of individual stakeholders, organisational practices and institutional bodies. It is increasingly being recognised by a wide body of academic and industry experts [6–16] that human behaviour has a significant role to play in reducing the environmental impact of the built environment.

Design decisions, material specification, working practices, building operation, user behaviour and legislative drivers can all combine with traditional, new and developing technologies to achieve low or zero carbon buildings which enable low environmental impact working practices and lifestyles, especially as modern living and working habits result in us spending a great deal of our lives within buildings.

Buildings and human behaviour are inextricably linked to environmental issues such as global warming and climate change, national environmental issues such as extreme weather events, the use and supply of energy, materials and water and local environmental issues such as water, ground and air pollution, localised flooding, high levels of water consumption and habitat destruction.

In essence this book focuses on the dynamic inter-relationship and synthesis of three disciplines, namely pedagogy, architecture and sustainability, as illustrated in Fig. 1.1.

The aim is to establish best practice in encouraging sustainable behaviour through the built environment by investigating both theory and practice of exemplar sustainable buildings through literature review, case study, interviews, surveys and questionnaires by engaging with key stakeholders such as developers, architects, engineers, contractors, building managers, teachers, curriculum developers and occupants.

The buildings used as case studies for this book, as presented in Chap. 5, attempt to achieve the lowest carbon emissions throughout their lifecycles by adopting low carbon design principles, sustainable materials and technologies, low environmental impact construction practices as well as considering the behaviour of building operators and users to optimise the ecological, social and economic potential of the buildings' sustainable features. Each of the buildings selected has a pedagogical function and the respective organisations strive to utilise the buildings as experiential educational tools to effect behavioural change beyond the buildings themselves.

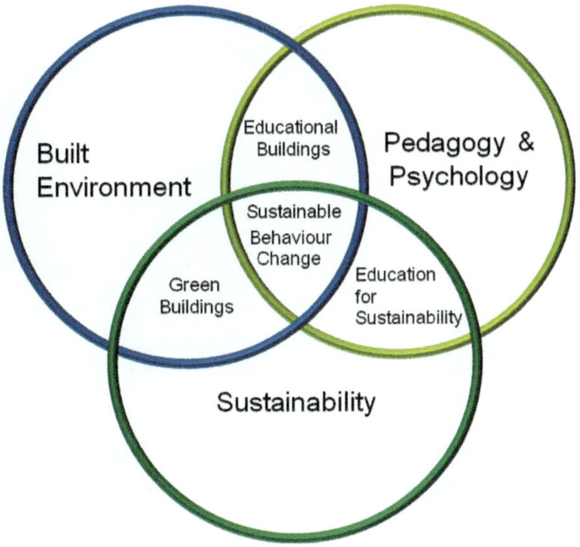

Fig. 1.1 Syntheses of three disciplines. *Source* Author

The book attempts to discover a link between buildings and human behaviour directly related to the goal of achieving greater sustainability both in the development of the buildings themselves and in the attitudes and behaviours of the stakeholders involved with the buildings, the intention being that the research will inform design and vice versa.

Significant progress has been achieved in the production of sustainable buildings from a technical standpoint and a good degree of work has been done in terms of the benefits of buildings on the health, well-being, performance and productivity of people and buildings. There is also strong body of knowledge and research in relation to the psychological barriers and drivers around sustainable and pro-environmental behaviour but little academic or industry work has been focussed on the inter-relationship of these key disciplines.

Over recent years there has been recognition from proponents of sustainable development of the gaps in knowledge within the field of sustainable buildings and how humans interact with them in order to achieve greater levels of environmental performance by balancing technical efficiency with behavioural efficacy.

This has been recognised by Sue Roaf, a leading academic and exponent of human-centred sustainable building design,

> We are now seeing an interesting new generation of research that links hard and soft sciences in a world where people cease to be the problem in low carbon buildings, but are recognised as a major part of the solution to achieving low carbon buildings in reality. [15], p. 7

This is supported and expanded upon by Hoffman and Henn [11] who state that obstacles faced by the green building movement are no longer primarily technological and economic, they are social and psychological (Fig. 1.2).

Fig. 1.2 MSc architecture students during the workshop at WISE building. *Source* Author

1.1 Sustainable Development and Architecture—A Brief History

The history of sustainable architecture is arguably the history of nearly all architecture and the evidence of the many historic buildings that survive to the present day is testament to this theory. Prior to the end of the eighteenth century the historic dominance of daylight, natural ventilation and organic materials had been the essential mode of environmental provision and all architecture, by this broad definition, could be considered as sustainable.

Most buildings employed the properties of material and form to make appropriate adaptations to the relationships between their uses and the surrounding climate. Their design was based on sophisticated empirical processes and relationships. These relationships were often codified in texts and treatises. The most familiar example being Vitruvius' Ten Books on Architecture expounding the virtues of *firmitas, utilitas and venistas*—solid, useful and beautiful. Vitruvius saw architecture as an imitation of nature, referring to the proportions of the human body, the 'Vitruvian Man,' as drawn by Leonardo Da Vinci (Fig. 1.3). The renaissance texts of Serlio and Palladio adapted similar principles to new and changing circumstances and this process continued in subsequent centuries.

The growth of what may be described, in the present context as 'unsustainable architecture' began at the end of the eighteenth century, with the industrial revolution, as new technologies were incorporated into the fabric of buildings primarily driven by new forms of energy and mechanisation. Residential, industrial and institutional buildings came increasingly to adopt mechanical systems of heating, ventilation and lighting in the quest to integrate the latest technologies driven by perceived social and economic benefits.

At its height the industrialisation of buildings produced the most resource-consuming structures, heavily reliant on energy intensive mechanical systems within sealed envelopes, with permanent artificial lighting and air conditioning. The subsequent history of this type of architecture correlates almost precisely with the historical

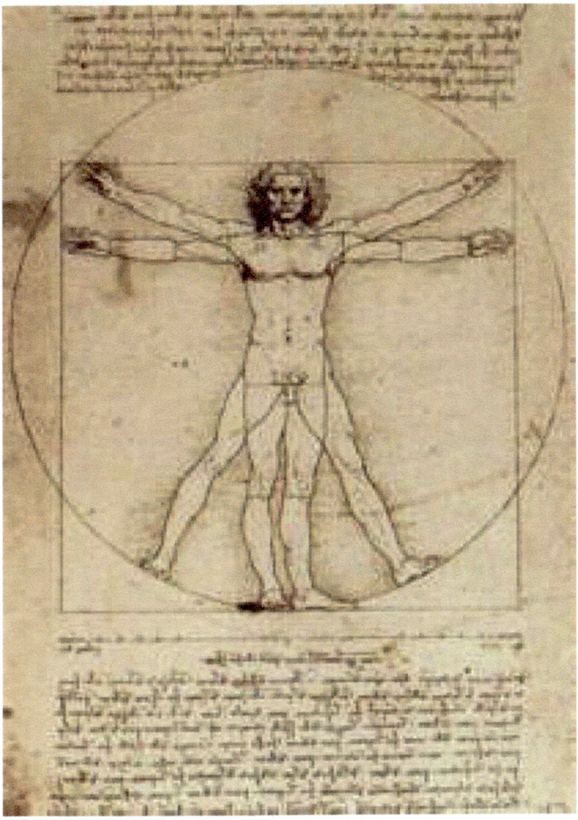

Fig. 1.3 Vitruvian man. *Source* azothgallery.com

increase in the global consumption of fossil fuels and the production of greenhouse gas emissions beginning the inexorable rise in global warming.

Arguably, the concept of 'sustainable architecture' would have been virtually inconceivable over fifty years ago because of a lack of appreciation for the environmental impact of buildings and their highly intensive resource use in both construction and performance. Around this time saw the re-emergence of the environmental nature of buildings. The emphasis was upon establishing a more deliberate link between buildings and the ambient environment.

It may be argued that the modern sustainable building movement grew out of the environmental concerns of the 1970s. The Oil Crisis of 1973 is often cited as the catalyst driving concerns of our over-reliance on natural resources. The architectural mantra of 'Long-Life, Low Energy, Loose-Fit' was coined by the then visionary RIBA President, Gordon [17] and was an early definition of sustainability for buildings, advocating the use of durable materials, energy-efficiency and adaptability.

As the movement has grown a substantial body of significant sustainable architecture now exists by a growing faction of designers, engineers and constructors. Five buildings representing leading edge sustainable design, construction and operation are presented in Chaps. 5, 6 And 7 of this study, supported by related design theory and practice from extensive literature review.

1.2 Defining Sustainable Buildings

Landman [18] suggests that a sustainable building represents building design and construction that uses methods, technologies, materials and processes that are resource efficient and that will not compromise the health of the environment or the associated health and well-being of the building's occupants, construction workers, the general public, or future generations.

According to the European Economic and Social Committee [19], p. 134 glossary of sustainability in the built environment sustainable construction is 'the application of sustainable development principles to the design and construction process i.e. use of fewer virgin materials, less energy in construction, less energy in use, less pollution and less waste; *whole life* approach to design, construction and life use; and providing safe places and work with acceptable social conditions integrated into sustainable communities.'

There are a number of international design standards developed over recent decades that have reached a high level of complexity in defining and assessing the environmental impact of buildings ranging from highly technical requirements including the UK Building Research Establishment Environmental Assessment Method (BREEAM) and the US Leadership in Energy Efficient Design (LEED) to more esoteric guides such as the Hannover Principles, Sim Van der Ryn and Stuart Cowan's Five Principles Of Ecological Design, the Todd's Principles of Ecological Design, the Sanborn principles, Eco-minimalism principles, Bioregional's One Planet Living principles and Dieter Ram's Ten Principles of good design.

The various principles of sustainable and ecological design focus on the interaction of architecture, products, people and nature. They take a broad perspective that incorporates cultural and historical traditions into the design process including human rights and sustainability, social aspects of sustainability and buildings, responsibility for the effect of design decisions and the use of nature as a model for design (biomimicry).

The internationally recognised Building Research Establishment Environmental Assessment Method (BREEAM) categories have been adopted as model benchmarks of sustainability as the basis for comparing and contrasting the selected buildings as case studies. The key sustainability indicators considered under the BREEAM method are:

1. Management
2. Health and wellbeing
3. Energy
4. Transport

5. Water
6. Materials
7. Waste
8. Land use and ecology
9. Pollution

However, even with this industry recognised standard there are many and widely varying barriers to achieving a sustainable building identified by Hoffman and Henn [11] as economic constraints, design limitations, wasteful construction practices, poor operation and profligate behaviour of those who use and operate buildings. The ultimate ambition of sustainable construction is to achieve buildings that have positive social, economic and environmental benefits, achieving zero carbon inputs and outputs throughout their lifecycles, even achieving carbon offsetting, acting as renewable energy generators, with 100% recycled and recyclable materials with zero waste.

It is difficult to establish a definitive description of sustainable building, architecture or construction, as a myriad of theoretical and philosophical perspectives exist. These include 'deep green' solutions that focus wholly on natural materials and systems and emphasise sensitivity and humility in relation to nature epitomised by the phrase 'to tread lightly upon the earth.' Other approaches include culturally based solutions that are highly contextual with forms, materials and construction methods echoing the local vernacular and emphasise local involvement and local expertise and the need for cultural sustainability. Technical solutions utilise science, economics and technology and generally emphasise global and generic i.e. non-vernacular expertise with leading edge contemporary international systems [20].

Sustainability has led to a revision of the way we view traditional or vernacular architecture. Rather than viewing it as retrograde or primitive, we now understand that it has something to teach us, that 'local' architecture grew out of many iterative attempts to deal with natural phenomena and should be respected as a 'repository of wisdom' [21], p. 22.

One of the most important elements of the sustainability movement has been the new attitude towards technology that it has encouraged. There is a growing debate today as to how best to tackle environmental degradation. The 'appropriate technology' movement, building on Schumacher's book *Small is Beautiful*, proposes modest solutions to environmental problems, claiming that high-tech answers just cause more problems and resource depletion. This approach has been dubbed eco-minimalism by the Director of Gaia Architects Howard Liddell [22] of the book *Eco-minimalism—the antidote to eco-bling*. The 'high-tech' advocates, on the other hand, continue to believe that all problems can be solved by science and technology [21].

This has led to a differentiation of sustainable building exponents, dubbed either as 'eco-centric environmentalists' or 'techno-centric environmentalists' by Pepper [23], p. 9. Guy and Farmer [24] propose that the ecological approach stems from a particular view of nature generated from an ecology-based conception, requiring a holistic framework of analysis emphasising the dynamic interaction between the living and the non-living as a community of interdependent parts. The issue extends beyond anthropocentric concerns to encompass a moral concern for the integrity of the natural world.

Guy and Farmer [24] suggest that approaches to building tend to draw directly on analogies from ecological systems as living, cyclical processes, which oppose the linear, open systems of conventional building. Ecological design strategies emphasise the reuse and recycling of materials and the reduction of dependency on infrastructure services of water, energy and waste through the use of appropriate technologies. In order to achieve this Pearson [25] asserts that a paradigm shift in society from mechanistic to a holistic systems-based conception of reality is needed, in other words a radical reconfiguration of values.

Examples of the eco-centric approach are exemplified by bioclimatic and biomimetic design principles. Bioclimatic architecture refers to the design of buildings and spaces based on local climate, aimed at providing thermal and visual comfort, making use of solar energy and other environmental sources. Basic elements of bioclimatic design, suggested by Olgyay [26], are passive solar systems which are incorporated into buildings and utlilise environmental sources (for example, sun, air, wind, vegetation, water, soil, sky) for heating, cooling and lighting the buildings.

Nature has evolved some of the most elegant and efficient design solutions since life on earth began. Biomimetic architecture gains inspiration from these designs and incorporates them into buildings which enable us to resolve design problems that complement the natural environment and satisfy our innate need for a deep connection to natural processes, commonly known as 'biophilia' (discussed in the next chapter). Victor Olgyay uses the term 'morphology in nature' by which the forces of nature have a direct effect on the formation of objects. He states that 'knowledge of form leads to the interpretation of the forces that moulded it and vice versa' [26], p. 62. Therefore, it can be argued that buildings should both be inspired by and reflect natural processes in their design, construction, operation and use.

The technological stance represents a belief in incremental, techno-economic change and that science and technology can provide the solutions to environmental problems. Cook and Golton [27] assert that techno-centrics recognise the existence of environmental problems and want to solve them through management of the environment. The design strategy is adaptive, but based on recognisably modern, usually high-technology buildings. The success of this design approach is expressed in the numerical reduction of building energy consumption and material embodied energy and in concepts such as life-cycle and cost–benefit analysis.

The techno-centric approach to sustainable building design is encapsulated in the maxim: *build tight, ventilate right*. An example of this approach is represented by the PassivHaus system. The term 'PassivHaus' refers to a voluntary, low-energy construction standard first developed in the early 1990s by Professor Wolfgang Feist of the PassivHaus Institute in Germany. The core focus of PassivHaus design is to dramatically reduce the requirement for space heating and cooling by super-insulating the building envelope, achieving high levels of air-tightness and using mechanical ventilation with heat recovery. Highly prescriptive requirements are set in order to achieve the PassivHaus standard in terms of overall energy required for space heating. Some aspects of bioclimatic design may be used, such as south-facing glazing, as well as external shading, but the overall build format and internal layout are not necessarily that different from those in a standard building.

In many respects sustainable building practices are an amalgamation of approaches, as the sustainable building agenda develops and becomes more synthesised with both traditional and modern methodologies. By studying these principles common themes emerge that can highlight opportunities which, if widely adopted, can encourage sustainable behavioural change throughout the planning, design and construction of buildings and beyond, in the way we operate and use buildings.

By embedding sustainable building techniques, interactive sustainable features, natural sustainable materials and renewable technologies, the building space becomes an example of sustainability in action throughout the design, construction, operation and use. This hypothesis proposes that this enables those involved throughout the building process from designers, engineers and contractors through to the building users and operators to directly experience and undergo 'deeper experiential learning' to ultimately offer solutions leading to more sustainable lifestyles and working practices. It is the aim of this study to establish the extent to which sustainable buildings can encourage greater sustainable behaviour and conversely how poorly performing buildings can discourage sustainable behaviour.

Aesthetics

The two different sustainable building typologies that represent both eco-centric (see Fig. 1.4) and techno-centric (see Fig. 1.5) approaches are aesthetically quite different from each other.

Sustainable design solutions that are based on natural, organic structures and forms, built from low-carbon materials, processes and technologies tend to appear more organic and non-linear in form than those that rely more on technical solutions to achieve narrower energy efficiency targets. These tend to appear more rectilinear in shape. This is of course a generalisation and as sustainable building design evolves, a synthesis of approaches is emerging. In terms of aesthetic appeal both have their supporters and critics and the discussion around the aesthetics of sustainable buildings is an increasing subject of critical debate.

It may be argued that sustainability, as a response to environmental problems, transforms the aesthetics of architecture in a substantive way. It may also be viewed as incompatible with pure aesthetic architectural design. This latter position is repudiated in the writings of Immanuel Kant:

> A work of free art (pure aesthetic consideration) does not possess an end other than to itself. Beauty is found in a works 'purposiveness' and the experience of beauty arise from the sense that a given object [e.g., a building] serves and fits a given purpose [e.g. sustainability].
>
> (Kant in Critique of Judgement [28], p. 47)

The eighteenth century philosopher who coined the term aesthetics, Alexander Gotlieb Baumgarten (in Healy [29], p. 22) describes aesthetics as 'a form of knowledge that is gained from that which is sensed.' Lee sums up Baumgarten's proposal:

Fig. 1.4 (Left) eco-centric sustainable building typology. *Source* CAT

> What we sense and perceive, the exteriority of an object, is a manifestation of the invisible or intangible qualities of its interiority, and therefore, that studying the connection of the two presents a meaningful approach to knowledge.
>
> (Lee [30], p. 11)

This strongly supports the hypothesis that sustainable buildings in both their form and function have the potential to influence knowledge, understanding and ultimately change the behaviour of those that are involved with them towards more sustainable behaviour. It may be argued that the external and internal appearance of buildings, constituted of the materials from which it is constructed, the method by which it is constructed and the processes that are evident in its operation have an impact on the behaviour of those involved in its design, construction, operation and use, inextricably linking form and function with a more whole-system approach.

Modernist design and construction practices have disconnected form from function in a more linear process with architects prioritising form and aesthetics whilst engineers, to a large extent, deal with function and performance afterwards. It is hard to imagine how the purely aesthetic considerations in the design of buildings can ever be achieved sustainably with much degree of environmental credibility. Perhaps one of the most notable examples is the London Bridge Tower (see Fig. 1.6), commonly known as 'The Shard,' undoubtedly a dramatic addition to the appearance of the London skyline, but in its scale and material considerations offers no relationship to its environment or climate, the aesthetic forcing it to be a climate-rejecting structure

Fig. 1.5 (Right) techno-centric sustainable building typology. *Source* Author

with completely internalised heating, cooling and ventilation systems. Its form has been criticised as aggressive and it tells us little about energy efficiency, prudent use of resources, equality and affordability.

In a recent article [31] the architect who led the team that designed the Swiss Re building (see Fig. 1.7), commonly known as the 'Gherkin,' Ken Shuttleworth, has pronounced that 'the age of bling is over.' He claims that tenants are demanding austere and efficient buildings that are more likely to be 'ground-scrapers' than high-rises. He has stated that 'the tall glass box is dead.' Despite this view, at a time when office rents are projected to surge, there is a boom in towering buildings and they are justified on the grounds of rising population and sustainability, but many developers still want to create 'stunning architecture,' and there will always be architecture driven by this desire, based mostly on aesthetic considerations.

Ken Yeang, a leading exponent of sustainable high-rise buildings asserts that the tall building is an urban fact. He seeks what he calls *ecomimesis* in buildings, a way to 'copy and paste' nature into high-rise designs. He states that:

> green high-rise buildings should not look like modernist buildings, not pristine and should have a level of 'fuzziness.' If we want eco-structures to be acceptable to the public they have to be aesthetically beautiful and the green aesthetic is something that we are constantly exploring. A building that acknowledges nature in form might help sharpen awareness about

Fig. 1.6 The 'Shard' buildings. *Source* Wikipedia 2013

> the role that architecture plays in our often un-green urban spaces. There is unquestionably a causal link between a building's look and its sustainable credentials.
>
> (Yeang in Pastemack [32], p. 2)

Natural buildings following the eco-centric design principles using natural materials, processes and technologies produce an aesthetic borne of sustainability-led thinking. These tend to reflect organic forms and natural processes that lead to a non-linear 'fuzzy' appearance which some find too far removed from the modernist aesthetic, not conforming to our aesthetic norms. Conversely, the techno-centric forms resulting from the application of sustainable technologies, such as the PassivHaus system draw criticism for their 'box-like' appearance and complete lack of aesthetic quality [20].

Many sustainable buildings seek to promote a green aesthetic and ambience in their design. Often these design considerations flow from and are achieved in a synergistic manner by the central design goals of reducing energy impact, water impact and providing a healthy inner space for its occupants. A green building can

Fig. 1.7 The 'Gherkin' buildings. *Source* Wikipedia 2013

aspire to and rise to a higher architectural realm and make a profound social/cultural comment, often achieving a remarkable level of beauty in the process.

Lee [30] argues that at the heart of architectural aesthetics are the concepts of (i) sustainability; a process that can be maintained and continued for a certain duration, or hypothetically speaking, indefinitely, and (ii) durability; the state of an object, the way an object is made allows it to function for the duration of the purpose it is intended to serve (and possibly beyond) without failing irreparably.

Hill [33] argues that the modern era has seen architecture increasingly participate in the endless search for new aesthetic directions and that current architecture is increasingly subject to 'aesthetic obsolescence' and suggests that sustainable architecture is susceptible to this pressure to become an 'aestheticised' commodity with an ever-decreasing life-span. Hill goes on to argue that the early radical environmental architecture of the 1960s and 1970s often focussed less on aesthetics and more on changing people's way of living (sustainable behaviour). However, he argues, more recent sustainable architecture has shown a greater interest in participating in what he calls the 'aesthetic economy,' rather than focussing on pro-environmental behaviour. It has instead focussed on developing technological strategies to maintain unsustainable behaviours for the lowest resource and energy cost.

Therefore, it is argued that what drives the design of a sustainable building ultimately impacts not just on its appearance but on its environmental impact through the behaviour of participants in the process. Technologies, materials and methods shape both the form and the function of the building and aesthetic considerations cannot be made in isolation.

1.3 Content of Chapters

A discussion of the case for sustainable buildings is the focus of Chap. 2 investigating global, UK and regional perspectives, current theoretical and philosophical thinking related to social, economic, environmental, technological, physiological and legislative issues.

Chapter 3 begins to draw together the inter-relationship between human behaviour, pedagogy, sustainability and the built environment defining sustainable behaviour from various psychological and theoretical perspectives, combining buildings and sustainable behaviour and considering buildings as educational resources. The chapter concludes by considering this synthesis in the context of individual, organisational and institutional behavioural change within the built environment arena.

The effects of human behaviour throughout the lifecycle of buildings is discussed in Chap. 4 in the context of achieving sustainable outcomes in the built environment and how these can be optimised through planning, design, construction, operation and use. Chapter 5 introduces the five buildings selected for case study research, describing their design origins, key sustainable features, methods and materials.

Chapter 6 brings together the theoretical findings and case study data from the online surveys and face-to-face interviews and discusses the results with analysis, restating the main research question and how this study aims to inform the design, construction, operation and use of buildings to achieve greater economic, social and environmental sustainability.

Chapter 7 highlights the key conclusions. It includes a summary of recommendations which is intended to be a useful reference guide for stakeholders in sustainable buildings for optimising the sustainable design, construction and operation of buildings in relation to sustainable behaviour.

Chapter 2
Defining the Need for Sustainable Buildings

2.1 Introduction

This chapter puts the case for a sustainable built environment in a global, national and regional context. It defines sustainable buildings from a number of theoretical, philosophical and typological perspectives that represent current thinking and best practice in the field. It critically analyses sustainability and the built environment on social, economic, environmental, technological and physiological grounds.

2.2 The Three Pillars of Sustainability

In 1987 the UN Environment Commission, chaired by Gro Harlem Brundtland, published its findings in *Our Common Future* which contains the widely accepted definition of sustainable development as: 'Development that meets the needs of the present without compromising the ability of future generations to meet their own needs.' [34], p. 10. This may be viewed as a virtuous but imprecise concept, open to various and often conflicting interpretations. However, it remains the global standard and addresses the needs of both the present and future generations in terms of environmental resources. The definition Brundtland coined may well be the single biggest imperative for global development in the twenty-first century. The Brundtland definition has spawned a series of sub-definitions to meet particular sector needs. These various definitions show the value of coining terms of reference for specific topics—be they building types, services provided, or levels of development which outline a philosophy that benefits from a degree of imprecision. There is general understanding and set of principles which allow useful sub-definitions to be framed within its broad meaning.

J. L. Clarke, *Sustainable Buildings for Sustainable Behaviour: Theory and Practice*, Green Energy and Technology, https://doi.org/10.1007/978-3-031-98905-6_2

Within these broad definitions and interpretations there are three recurring dimensions that provide the focus for action by different interested parties; environmental sustainability, economic sustainability and social sustainability.

The widely accepted three pillars of sustainability of economy, society and environment were discussed at the 2005 World Summit of the United Nations and are represented in Fig. 2.1 which suggests that both economy and society are constrained by environmental limits. Figure 2.2 shows sustainability at the confluence of the three constituent parts. The three pillars, or triple bottom line have served as a reference point for numerous sustainability standards and certification systems in recent years and have been adopted for this study as a baseline standard for analysing the sustainability of buildings.

Economic and social systems cannot be divorced from the 'carrying capacity' of the environment—the idea that growth and social welfare has to be balanced by the conservation of environmental resources by the present generation for the benefit of future generations. Hence the term 'sustainable development' has wide ramifications for built environment professionals—the people who are carrying out the development.

Sustainable development, defined in such terms, can be something that almost everyone can subscribe to without too much inconvenience and can be interpreted into a set of practices that are reasonably congenial in almost any socio-political context and therefore tends to reflect highly anthropocentric and economic motives that lead to nature being seen as a resource and is therefore open to criticism. It also begs the question of whether environmental and economic sustainability are truly reconcilable and whether built environment professionals are misleading themselves

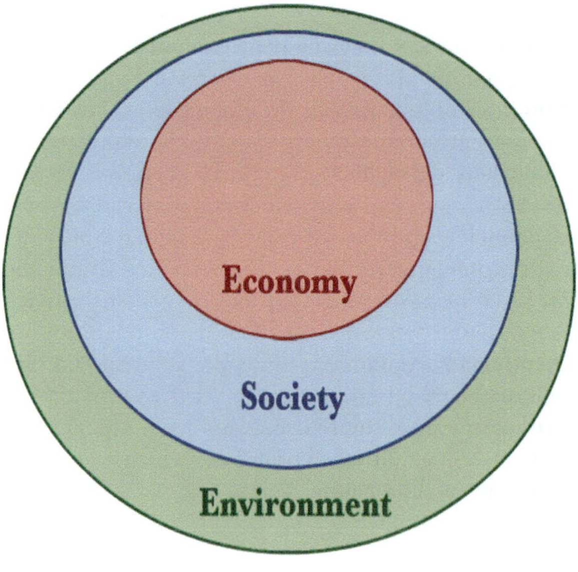

Fig. 2.1 The relationship between the three pillars of sustainability. *Source* Cato

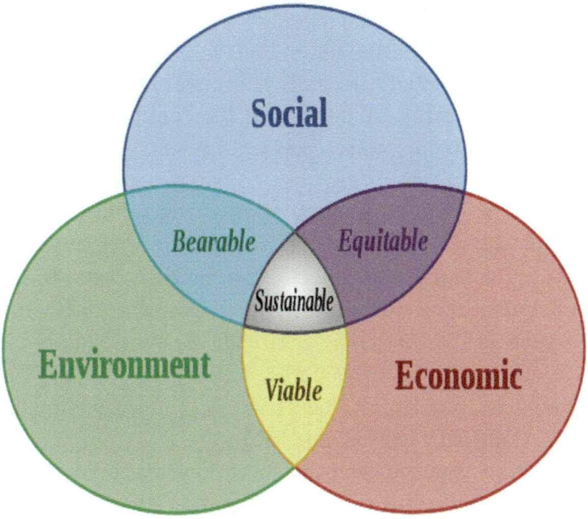

Fig. 2.2 Scheme of sustainable development. *Source* Adams

into thinking that development can ever be completely sustainable in a world of finite resources.

2.3 Environmental Perspectives

The Global Perspective

Anthropogenic, human-induced global warming through the greenhouse effect has largely been recognised by internationally accepted authorities on climate change, most notably the Intergovernmental Panel on Climate Change (IPCC) to be significantly contributing to a rise in global temperatures with resulting changes in climate stating that 'the balance of evidence suggests a discernible human influence on global climate'.

A small rise in overall temperature means that water stored in icecaps and glaciers begin to melt, so sea levels rise and the reduction in ice cover decreases the ability of the Earth to reflect heat back into the atmosphere. Warmer temperatures mean that water vapour evaporates from oceans, so more moisture is available for rain, and the presence of excess moisture in the air contributes to shifting wind and weather patterns. This has far-reaching global implications as we experience flooding of coastal and low-lying areas, heat waves become more frequent and more intense, drought and wildfires are likely to occur more often. Disease-carrying mosquitoes may expand their range, Eco-systems and food production will be disrupted and species may be pushed to extinction.

Figure 2.3 illustrates earth-surface temperature rise and shows that the average Earth's land temperature has warmed by 1.5 °C over the past 250 years and that the temperature rise correlates closely with CO_2 concentrations and the period of increasing industrialization and development of carbon-based economies, which can be largely attributed to the burning of fossil fuels. The term 'global warming' suggests an even heating of the Earth; the reality is that there is a great deal of regional instability with some areas warming and getting wetter and some cooling and getting drier.

At the time of writing, it has been widely reported that the area of Arctic and Antarctic sea ice has shrunk significantly against previous records [35] as shown in Figs. 2.4 and 2.5. Sea ice in the Arctic is seen as a key indicator of global climate change because of its sensitivity to warming and its role in amplifying climate change. Sea ice is known to play a critical role in regulating climate, acting as a large reflector of much of the sun's energy, helping to cool the Earth. According to the National Snow and Ice Data Center (NSIDC), the warming of Arctic areas is now increasing at around 10% per decade [36].

Arctic sea ice follows an annual cycle of melting through the warm summer months and refreezing in the winter. Other leading ice scientists have predicted the complete collapse of sea ice in the Arctic within four years. 'The final collapse is now happening and will probably be complete by 2015/2016,' according to Professor Peter Wadhams of Cambridge University (in [35], p. 2).

The design, construction, operation and use of buildings has a significant impact on the global environment in terms of energy use, the burning of fossil fuels, CO_2 emissions, the depletion of finite resources, air, water and ground pollution, biodiversity and the production of waste on a global scale. Near the end of the twentieth century, the built environment became a focus of attention within the global environmental movement. Research by the World Green Building Council reveals that buildings

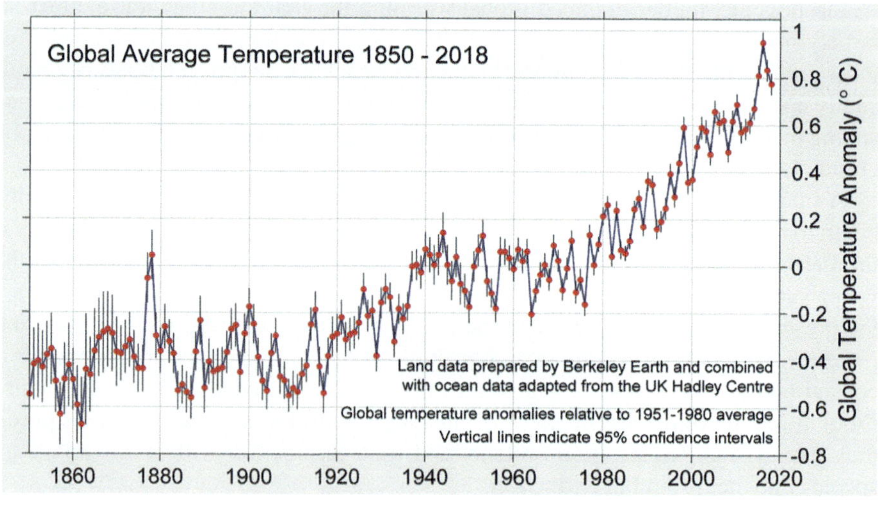

Fig. 2.3 Earth-surface temperature rise. *Source* www.berkeleyearth.com

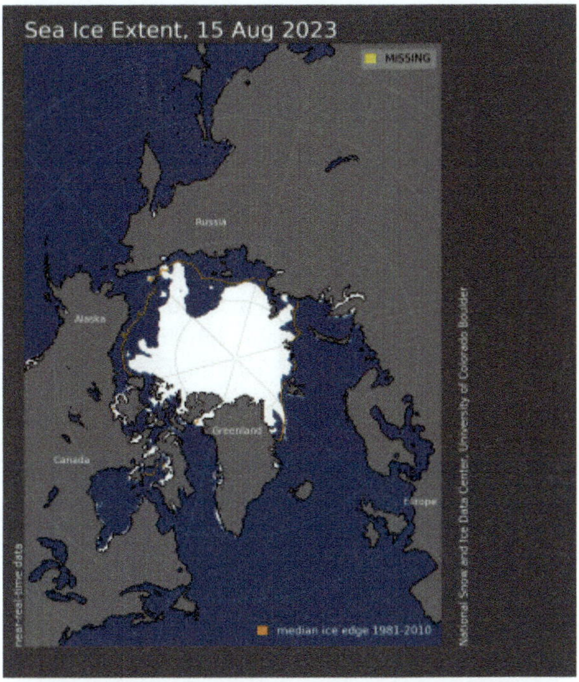

Fig. 2.4 Arctic sea ice extent for August 15, 2023, was 5.74 million square kilometres (2.22 million square miles). The orange line shows the 1981–2010 average extent for that day

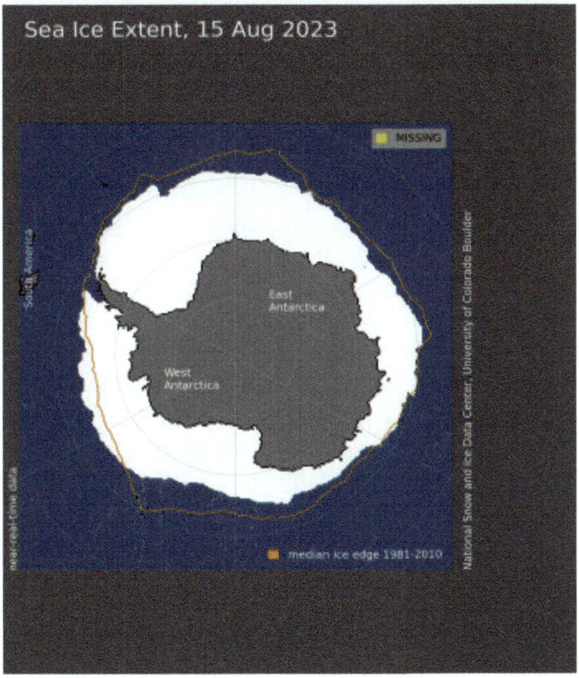

Fig. 2.5 Antarctic sea ice extent for August 15, 2023 was 15.12 million square kilometres (5.84 million square miles). The orange line shows the 1981–2010 average extent for that day

consume 40–50% of the world's materials, use a quarter of the wood harvested globally, are responsible for about 20% of global water consumption, consume between 30 and 40% of the world's energy and create one quarter of total world carbon emissions that cause global warming.

In spite of international agreements (such as those signed at the Rio Earth Summit, Kyoto and Paris) CO_2 emissions continue to rise, around a third of which is due to global built environment and related infrastructure activity, and that in spite of efforts made in advances to improve the sustainability of the built environment. There are a number of reasons for this, the dramatic increase in human population anticipated to exceed 9 billion by 2050 [37], the legacy of older inefficient buildings, continuing use of unsustainable building practices and unsustainable behaviour in the way we use and operate buildings, rising consumer standards and demand for energy intensive goods and rising demand for non-renewable energy from newly industrialised countries (Brazil, China and India).

It was argued by Dunlap and Van Liere [38, 39] that a fundamental shift in our world view is needed from a dominant Western view that sees humans as separate from nature with dominion over all other organisms, that we are masters of our own destiny and have the intellectual and technological capacity to solve any problem, that we have access to an infinite amount of resources to a new ecological paradigm in which humans are interdependent with other organisms. We need to acknowledge that many things we do have unintended negative environmental consequences for the environment, recognition that many resources such as fossil fuels are finite and that ecological constraints, such as the carrying capacity of an environment, are placed upon us. This argument can be viewed in the context of buildings and is a strong argument for sustainable buildings that have a positive effect on our behaviour and the environment.

There is strong international consensus that climate change is happening and the vast majority of people, particularly in the developed world, claim to be concerned [40]. But efforts to communicate or convince still seem to fail to result in global action as evidenced by the rise in global temperatures (see Fig. 2.3), CO_2 emissions and indicators of unsustainable behaviour such as the exponential rise in car ownership [41]. So why then are we not adopting sustainable solutions faster and more widely to tackle the problems of climate change? Climate change denial, apathy and scepticism are complex psychological, societal, emotional and political problems.

A useful comparison can be made with mass human rights abuses. Cohen [42] shows how people living under repressive regimes resolve the conflict between the moral imperative to intervene and the need to protect themselves and their families. He found that people deliberately maintain a level of ignorance so that they can claim they know less than they do. They exaggerate their own powerlessness and wait indefinitely for someone else to act first—a phenomenon that psychologists call the 'passive bystander effect.' Both human behaviour strategies lie below the surface of most of the commonly held attitudes to climate change. Cohen [42] also observes that societies also negotiate collective strategies to avoid action.

Dr. Kari Marie Norgaard of the University of California [43] reaches a similar conclusion and argues that denial of global warming is socially constructed and that climate change is excluded from social norms what she calls. According to Norgaard most people have tacitly agreed that it is socially inappropriate to pay attention to climate change, through a selective framing that creates the maximum distance. In opinion poll research [44] the majority of people will define the problem in three ways; (i) as being far away ('it's a global problem, not a local problem') or far in the future ('it's a huge problem for future generations'), (ii) with false positivity ('technology will solve the problem') or ('the markets will sort it out') and (iii) reactive denial or projection ('I've got enough problems') or ['it's not me, it's those other people…' (the rich, the poor, the Chinese)].

Marshall [44] puts forward reasons why we deny climate change (Table 2.1).

Marshall [44] goes on to offer solutions:

- We recognise that information alone cannot produce change.
- We openly recognise the tendency to denial.
- We encourage emotional responses.
- We develop a culture of engagement that is visible, immediate and urgent.
- As individuals as part of society we act with integrity and clarity.

Causes of climate change have been bundled with global 'environmental issues' dealt with by environment ministers and departments and talked about in the media by environmental reporters. The issues are championed by environmental campaigners and therefore filter the message through a minority ideology or world view. Thus, the issue has been burdened with a set of associations and metaphors that allow the general public to exclude it from their primary concerns ('I'm not an environmentalist') as could senior politicians ('environment is important but economy and jobs are my priority'), in other words outside their norm of attention.

Climate change is invariably presented as an overwhelming threat requiring unprecedented restraint, sacrifice and government intervention. The metaphors that it evokes, such as 'live simply so that others may simply live' are anachronous to people who feel rewarded by free market capitalism and distant government interference.

Phrases such as 'save the planet' have little or no emotional connection for most people. The word 'save' is closely connected with campaigns such as 'save the whale' with connotations of activist culture and does not inherently require a change

Table 2.1 Reasons why we deny climate change

Our response is strongest to threats that are	Climate change is
• Visible	• Invisible
• With historical precedent	• Unprecedented
• Immediate	• Drawn out
• With simple causality	• A result of complex causes
• Caused by another 'tribe'	• Caused by all of us
• Direct personal impacts	• Unpredictable and has indirect personal impacts

in behaviour and also speaks of sacrifice, struggle and abstinence, to give something up, like heating or lighting. Experts in motivational theory say this is an unproductive way to change behaviour and it is better to focus on benefits. The word 'planet' like 'climate' has a technical scientific origin that distances it from everyday lifestyles.

Climate change is a fast-moving phenomenon and increasingly, severe climate events will reinforce the theoretical warnings of scientists with far more tangible and immediate evidence.

People will only accept a challenging message if it is conveyed in their own language and values which comes from a trusted communicator, usually within their own peer group. We need to recognize that people are best motivated by a positive vision, by understanding the real and personal benefits that could come from living in a low carbon world where buildings in entire neighbourhoods are being constructed and refurbished sustainably.

In an article by Brown he asserts that:

> The kind of changes in consumption needed to make a real difference to our carbon output would require multinational action at governmental level. But democratic governments act from perceived self-interest even more than individual voters do. Both ignore the very painful adjustments that would be needed to diminish climate change. They want to minimise their own unpopularity and will see the world in ways that make their actions seem rational. [45], p. 2

Even the Kyoto Protocol can be seen as an agreement that has already licenced wealthy countries to maintain their lifestyles without making any meaningful cuts in emissions or undertaking related policies [46]. The 2009 United Nations Climate Change Conference in Copenhagen resulted in no mandatory targets for emissions reductions.

The UK Perspective

The built environment in the UK accounts for between 40 and 50% of all CO_2 emissions amounting to 360 million tonnes of CO_2 released into the atmosphere per year [3] and produces 120 million tonnes of waste [47]. Studies have shown [48] that up to 29% of all solid waste going to landfill originates from construction and demolition. During the process of decomposition this can result in the release of methane, a greenhouse gas, with four times the global warming potential of CO_2.

There is an ever-increasing momentum for the provision of built environments as the UK population is expected to increase from 62.2 to 77 m by 2050 [49] and there is strong scientific consensus that human endeavours, such as construction, will contribute exponentially to global warming with resultant effects on climate change and sea level rises, with potentially disastrous effects for life on earth, unless fundamental shifts occur in the design, construction, operation and use of the built environment.

The Climate Change Act of 2008 sets legally binding national greenhouse gas reduction targets of 34% by 2020 and at least 80% by 2050 compared to 1990 levels. There have been a number of initiatives over the past few decades focussed on reducing the environmental impact of the built environment. The recognition that building practices need to change were evidenced by revisions in legislation with targets for zero carbon and low water usage new-build housing by 2016 and new build non-domestic buildings by 2019 for England and Wales [50].

The UK Government's first national Climate Change Risk Assessment [51] highlights UK threats from global warming under current conditions and over the long term, within three timeframes: 'the 2020s' (2010–2039), 'the 2050s' (2040–2069) and 'the 2080s' (2070–2099).

The assessment acknowledges that the climate is changing (see Fig. 2.6) and warming will continue over the next century and the UK is already vulnerable to extreme weather, including flooding and heat waves. Increases in winter rainfall are predicted at around 16%, increasing the risk of flooding from rises in sea and river levels by up to 36 cm and more days of intense rainfall. A decrease in summer rainfall of around 22% are predicted with implications for water use restrictions and long-term shortages and even drought.

Warmer and drier summers, up by 2 °C will potentially increase health threats affecting the most vulnerable groups in our societies. Sensitive ecosystems are likely to come under increasing pressure and climate risks in other parts of the world could have a significant indirect impact on the UK with threats to global health, political stability and international supply chains.

The report highlights, among other issues, buildings and infrastructure in the urban environment as potential risks due to climate change that require 'early action.' Energy demands for cooling are already increasing and are likely to increase further, particularly in the south of the UK and we are beginning to see a significant number of dwellings with cooling capacity [52]. Overheating is projected to pose an increased risk to building occupants. Summer overheating is projected to emerge as a significant risk, potentially contributing to health-related problems. This is now being addressed

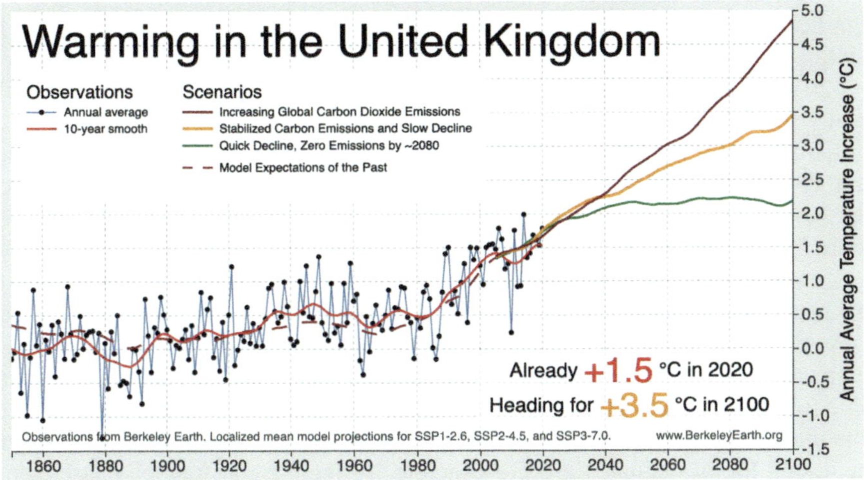

Fig. 2.6 Warming in the United Kingdom. *Source* www.berkeleyearth.com

as a mainstream concern within the Building Regulations (Approved Document Part L) for new buildings.

UK water resources are projected to come under increased pressure with more frequent water use restrictions and, in the long term, water shortages. By the 2050's between 27 and 59 million people in the UK may be living in areas affected by water supply–demand deficits [51]. Flood risk is projected to increase significantly across the UK and annual damage to UK buildings due to flooding from sea and rivers currently totals around £1.3 billion and is projected to rise to between £2.1 billion and £22 billion by the 2080's [5]. Without action, a range of important infrastructure such as roads and railways could be affected by a significantly increased risk of flooding.

Energy demands for cooling are likely to increase, particularly in the south of the UK. Currently cooling of buildings including air conditioning, refrigeration and cooling of information and communications technology infrastructure accounts for around 4% of total UK electricity use and demand for cooling of buildings is already increasing. In addition, the waste heat from the cooling is typically ejected into the environment, rather than captured, recovered or reused.

Flood risks to buildings and key infrastructure are anticipated to increase. The Urban Heat Island Effect could become more pronounced. Large cities in the UK already experience higher night-time temperatures than the surrounding countryside due to their absorption of heat during the day. Increasing urbanisation and energy use would cause this phenomenon to become even more noticeable over the course of the coming century, exacerbating potential health problems and impacts on biodiversity caused by overheating. By the 2050's London temperatures could exceed the official definition of a heat wave (32 °C by day, 18 °C by night) for more than a third of the summer.

Sewers are projected to fill more frequently and spill into rivers and the sea. Although heavily influenced by socio-economic factors such as population growth, significant increases in spill frequency may occur in the future due to changes in rainfall patterns and may impact public health and biodiversity.

420 million tonnes of materials are used in the construction of buildings in the UK each year [53] which accounts for 30–50% by volume of all manufactured goods, excluding food production [54]. 120 million tonnes ends up as waste from construction, demolition refurbishment and excavation processes and it has been estimated that 20 million tonnes of *unused* materials end up in landfill each year [47]. Over their entire lifecycle, materials used in construction contribute a significant amount to the environmental impact of the construction sector in terms of the extraction of finite raw materials, their processing, transportation, manufacture into building products, packaging, installation on-site and their destination after primary use.

In order to reduce the environmental impact of building materials those responsible for creating and maintaining our built environment must engage with sustainable material issues at each stage of procurement, design, construction and beyond the design life of the materials. For example, during the construction phase alone, research has shown that CO_2 emissions can be reduced by as much as 30% through a careful selection of low environmental impact materials [55].

The issue needs to be addressed across the whole construction cycle from the brief through to construction of the project and beyond the life of the building. This has led to the development of a range of analytical techniques collectively called environmental life-cycle assessment (LCA). There are also number of guides, resources and initiatives aimed at tackling the environmental impact of construction materials, including the Building Research Establishment (BRE) Green Guide to Specification and the Waste and Resources Action Plan (WRAP) 'halving waste to landfill' initiative.

The report concludes that despite the uncertainties to future climate change and its impacts, the evidence is now sufficient to identify a range of possible outcomes that can inform adaptation policies and planning. Lord Krebs, chairman of the group advising the Government on adapting to climate change, said 'without an effective plan to prepare for the risks from climate change the country may sleepwalk into disaster' [56], p. 22.

Adaptation and Mitigation in Relation to Climate Change

Mitigation of climate change deals with the impact the built environment has upon climate change. Roughly half of man-made carbon emissions result from buildings and when transport is included the figure approaches 75% [57]. The challenge is to act to try to limit future climate change by reducing the carbon emissions that are understood to be its root cause, largely from the burning of fossil fuels in the embodied energy from construction materials, the construction process and from buildings in use. Adaptation recognises that climate change is inevitable and is concerned with how design practices need to be changed to adapt to a warming world.

This requires a fundamental shift in built environment practices and the behavioural change of individuals, organisations, national and international institutions, away from past experiences that were related to a familiar and relatively predictable climate. The design, construction, operation and use of buildings in the modern-era has increasingly displayed significant levels of CO_2 emissions [3], relying on energy-intensive technological solutions and practices far removed from natural and passive systems that are now proving to be unsustainable and exacerbating climate change.

Microclimates can be modified to enhance human comfort levels, using less energy. Cold northern cities can benefit by improving insulation levels and harvesting solar radiation, both relatively simple and cost-effective solutions. In hot climates, solar design and shading devices, including trees and climbing plants, can be used to create shade and the cross-section of buildings can be altered to channel natural air currents, thereby reducing the need for energy intensive air conditioning.

2.4 Social Perspectives

Modern European societies are characterised by the Cartesian trademarks of dualism, reductionism and positivism which typically shape the way we think about problems, make decisions and therefore the way we design buildings. Dualism expresses a difference between body and mind, matter and spirit, and between reason and emotion by which all phenomena can be determined by mechanistic and technological principles. This enables the separation of regular and controllable events from those that are erratic, unpredictable and uncertain, such as the consequences of anthropogenic climate change. Cartesian dualism effectively sets humans apart from nature and individuals apart from outside influences. Responsibility for outside influences is dealt with by articulating codes of appropriate behaviour, morals and ethics.

Architecture has increasingly become a science-based discipline, relying heavily on technical solutions, and has therefore, according to Dripps [58] become separate by disconnecting people from its description of the world. By definition, reason-determined solutions become the only consideration and the emotions of individuals are constrained and moral sentiments are excluded from the process, making a reliance on technical and scientific solutions the norm.

Reductionism perceives all entities as consisting of simpler or more basic entities, breaking down a problem into simpler units, its component parts. We study and attempt to understand these simple units by reassembling the parts in a logical fashion to understand the whole problem. The whole consists of the sum of its parts, no more and no less [20]. Reliance on this system is the trademark of positivism, the belief in the infinite capacity of human reason to control, dominate, and put to work the forces of nature [59].

This has led to the familiar distinction between the 'art of architecture,' the humanistic components of social, moral and ethical considerations and the 'science of architecture,' the mechanistic, technological expression of architecture that has prevailed throughout the industrial revolution. Considering the world as systems to be exploited and manipulated for human purposes has resulted in the destruction and pollution of much of the natural environment. Since the 1970's 745,289 km^2 of Amazonian rainforest have been destroyed [60] mostly due to illegal logging. Since the industrial revolution acidity in the oceans has increased by 30% [61] and rigorous scientific debate is anticipating a sixth period of global mass species extinction, due in large part to human activity; the intensification of agriculture, increasing levels of human settlement, industrial pollution and the accumulation of greenhouse gases [62].

According to Williamson et al. [20] science has become one of the most influential ways of understanding the world. Society displays confidence in scientific methodology that has driven industrialisation, leading to new technologies that have undoubtedly contributed to material well-being and health for some parts of the world. It has also contributed to the exploitation of limited global resources due to industrial-scale production and consumer demand for processed goods, impacted on biodiversity through the wide scale extraction of materials from the land and oceans and adversely affected the health and well-being of large populations through air, water and ground system pollution.

Williamson et al. [20] assert that modernity continues as the dominant framework of architectural practice, as manifest in its political context, legislation, regulations, design advice, and other practices, 'post-modern' theorists and critics point to the enormity of the predicaments we face and repudiate the modern ways of going about solving the environmental, economic and social problems.

A more ecological world view is emerging from the prevailing scientific paradigm that derives from the notions of ecology as the new science of the relationship between organisms and their environment, or the relationship between a human group and its environment. Ecology provides insights about how natural systems work, including systems subject to human interference, such as the built environment, and provides a scientific justification for sustainability but does not generally consider the more complex issues of human behaviour and interventions (discussed in Chap. 3).

In a reaction to the lack of scientific consensus there is a strong argument for adopting the 'precautionary principle.' The following statement summarises the principle as:

> When an activity raises threats of harm to the environment or human health, precautionary measures should be taken even if some cause-and-effect relationships are not fully established scientifically.
>
> (The Wingspread Statement [63], p. 3)

The principle implies that there is a social duty to act when scientific investigation has found a plausible risk. The application of the precautionary principle in modern policy making is hampered by political will, the wide range of interpretations placed on it and the often-irreducible conflict between different interests, so that the debate becomes highly politicised.

The humanities in the age of technology are often seen as a separate resisting force against the growing domination of science and technology but there is common ground that requires collaboration across all spectrums. Architecture is a practice that demands negotiation and dialogue between practical, theoretical, social, political, cultural and technical concerns that contribute towards the development of design ideas and their translation into built form.

This broadly ethical dimension of the humanities applied to architecture links the built environment disciplines with social cohesion, moral imperatives and ethical behaviour which can be directly linked to the need for sustainable architecture. It may be argued that the humanistic basis for architecture, prevalent for centuries has been gradually eroded and marginalised in contemporary society, evidenced by the teaching and researching in humanities-based subjects in universities today, and conversely the rise of their more accountable counterparts in science and technology. This has also brought about a growing emphasis on specialisation at the expense of interdisciplinary work characteristic of the humanities. With the built environment disciplines we are in danger of giving priority to short-lived concerns of novelty and visual appeal at the cost of issues of meaning, such as climate change and the sustainability of the built environment.

> When designers get it right, they create in a way that reinforces our common humanity at the deepest level.
>
> (Orr [14], p.24)

Orr [14] goes on to assert that designers ought to think of themselves as place-makers, not merely as form-makers. Form-making puts a premium on artistry and fashion. It is mostly indifferent to human and ecological costs incurred elsewhere. The first rule of place-making is to honour and preserve other places in terms of both space and culture.

Hajer [64] argues that to analyse environmental questions in terms of quasi-technical decision-making on well-defined physical issues misses the essentially social questions that are implicated in the details. He goes on to suggest that the concept of sustainable building is fundamentally a social construct and, in order to understand sustainable buildings more fully, we have to account for the social structuring of both the identification of environmental problems and their resulting embodiment in built forms.

Understanding the concept of a sustainable building as a social construct does not seek to deny that there are serious environmental problems in urgent need of addressing or that the wide range of built responses are not valid ethically, socially, commercially or technically, in their own terms. The premise is that individuals, groups and institutions embody widely differing perceptions of what environmental innovation is about. Sustainable buildings can be viewed as a social representation of differing ecological and ethical values encompassing a broad range of issues with differing emphasis on technique, aesthetics and social responsibilities. One of the questions tackled in Chap. 3 is why we often act in the self-interest, often in terms of individual economic and health concerns when we often fail to act on wider concerns of community, social cohesion, education and public health.

Many UK cities are now entering an era when population expansion and housing shortages may well exacerbate existing social inequalities and cause huge infrastructure pressures.

Professor Tim Dixon of the Oxford Institute for Sustainable Development defines social sustainability in his paper 'Putting the 'S-Word' Back into Sustainability,

> How individuals, communities and societies live with each other.... blending traditional social policy areas and principles, such as equity and health, with emerging issues concerning participation, needs, social capital, the economy, the environment, and more recently with notions of happiness, well-being and quality of life.
>
> (Dixon [49], p. 4)

Dixon highlights ways of measuring social sustainability and identifies the operational issues in mainstreaming social sustainability. The paper highlights that social sustainability can operate across different scales: it can operate at a business level in terms of the way in which an organisation engages with society, e.g. corporate social responsibility; it can operate at an individual building level through the way in which the building is connected to the wider community or neighbourhood and as the aggregate of the relationships between people and places within a community or neighbourhood.

Factors which inhibit change are: the limitations of the building industry to measure social sustainability. Metrics are relatively poorly developed (in comparison with, for example environmental sustainability) at corporate level and at an individual building level where community and impact-based measures are weakly formed such as with BREEAM, and at community level. Where measures for social impact are relatively well-developed these tend to be aspirational and often not followed through or monitored over time. Developers, with few notable exceptions, have no long-term interest in monitoring the success or otherwise of the project in social terms.

Given the impact of the built environment on the natural environment, society and culture it can be strongly argued that the built environment professionals have a moral and ethical imperative to act as stewards or trustees, to pass on the best of our civilisation (a stable climate and biological diversity) to future generations.

The responsibility of the current generation of developers, architects, engineers and constructors to act is arguably greater than any that have preceded them and the dereliction to act could be severe for the global environment, the world economy and all of society.

The effects of climate change are now believed by a growing consensus of world scientific knowledge to have very serious implications for the way our climate works and the moral and ethical imperative to act is clear but much of the current debate around climate change focuses on disputes over historical temperature records, the likely rate of warming or the viability of a carbon tax, and the extent of the ethical challenge can easily be overlooked or marginalised.

In his article *A Perfect Moral Storm* Stephen Gardiner states that 'prosperous people today are knowingly imposing large burdens on poor people, on the biosphere and on future generations, and knowingly failing to act to reform our collective behaviour' (Gardiner [65], p. 2). He suggests that if we measure our behaviour in terms of carbon-dioxide emissions, today's prosperous people are not even beginning to reform their/our conduct and points out that we are emitting more than we were 20 years ago. Global carbon emissions from energy production are up 48% on 1992, when the original Earth Summit took place in Rio [66].

Strong ethical claims do not always get transformed into strong political positions despite how convincingly the ethical argument is put, the practical impacts remain to be demonstrated while emissions from carbon uses that benefit the developed world are already making life worse for poor people due to climate change, the present generation continues to cause harm in the future.

Typical professional moral behaviour is expressed in terms of guidelines, rules, standards and codes. As Tom Spector observes; 'the nature of building codes reinforces the idea that professional moral obligations exist within a framework of well-defined relationships, expectations and activities' [67], p. 22. Steele expands upon this by stating 'the new ecological awareness requires architects to be more open to other fields and to read more widely in socio-economic issues, so that more informed decisions can be made. This is not easy in a profession that can be very insular and focused' [21], p. 7.

Relying on this conventional mode of responsible decision-making (behaviour) to achieve a truly sustainable architecture is problematic. The construction and development industries are heavily dependent on the economic and political systems within which they operate. Ethics and codes of conduct become secondary within a free-market economy. Decisions need to extend beyond a reliance on existing conventions and empirical knowledge and will require strategies over and above legal obligations of building standards and regulations. Some of the best examples of sustainable building projects are already achieving this and five of them are presented later on in this thesis as case studies. In the words of the architect for one of the case study buildings 'the ethic is rendered aesthetic.'

2.5 Physiological Perspectives

There are health threats related to climate change, affecting the most vulnerable groups in our society. These are likely to place different burdens on public health and social care services. Premature deaths due to hotter summers are projected to increase by between 580 and 5900 by the 2050's [51] and the resilience of our built environment has a large part to play in adapting to and mitigating for climatic change to provide conditions for humans to function effectively and comfortably in a warming world.

Generally, modern and industrialised construction methods, materials and technologies have increasingly produced buildings that have detrimental effects on occupant health and well-being as witnessed by the recognition of various medical conditions and disorders such as Sick Building Syndrome (building-related sickness), Seasonal Affective Disorder and Nature-deficit disorder.

Sick Building Syndrome

Sick Building Syndrome (SBS) is a chemical, biological and psychological phenomenon commonly defined by symptoms associated by being in a particular building which are relieved by leaving that building. It is a general malaise of multiple symptoms of an unknown or unclearly recognised cause. It should not be confused with 'building related illness' (BRI) in which both the disease and its cause is known—such as legionella, asbestosis, or humidifier fever.

Typical symptoms include headaches, fatigue, lethargy, eye irritation, nasal irritation, dry throat, loss of concentration and nausea.

Building features that are strongly associated with symptoms of sick building syndrome are highlighted by Bradley [68] and summarised below.

- A hermetically sealed, airtight shell
- Mechanical heating, ventilation and air conditioning systems (HVAC)
- Use of materials and equipment that give off a variety of irritating and sometimes toxic fumes and/or dust (volatile organic compounds, bacteria, moulds etc.)
- Fluorescent lighting that may produce photochemical smog

- Application of (crude) energy conservation measures
- Lack of individual control over environmental conditions
- Electromagnetic fields
- Negative ion depletion

The reasons for these failings are often integrated at design stage, such as airtight envelopes with inadequate ventilation and can be related to the narrow parameters under which buildings are operated in relation to temperature, relative humidity, lighting levels and noise criteria. These parameters are often operated at default settings which do not allow for fluctuations and human variability. As buildings become more complex, these problems can be exacerbated if operational systems are poorly commissioned and maintained. Professor Sue Roaf states;

> The widely held assumption that the health, well-being and comfort of building occupants depend on the close control of indoor climates is...erroneous and that regulations that posit this view promote the values of certain professional and cultural groups and ignore local cultural, economic, climatic and environmental factors.
>
> (Roaf et al. [69], p. 65)

Even those buildings that have sustainability as a key design intent can overlook the health and well-being impacts of sustainable design decisions. Ironically, 'passive' techniques that are considered highly sustainable such as solar design and airtightness can result in overheating and poor indoor air quality if they are applied without proper consideration of their impact on the health and well-being of occupants or the ability to operate and maintain the systems at optimal levels through the life of the building.

Building techniques that allow the building to 'breathe' alleviate a lot of problems with indoor air quality but are often not understood by building financiers or poorly executed by designers, engineers and constructors. The notion of the building as a 'third skin' comes from the 'Building Biology' movement: the 'first skin' being the human skin with clothes forming the second skin.

A New Economics Foundation report [70] suggests three core aspects of well-being to which the built environment can contribute:

- Personal well-being based on people's experience of life in relation to their physical and psychological well-being;
- Social well-being focussed on people's experience of life in relation to their community; and
- Economic and material well-being based on people's life experience in relation to conditions and circumstances and their physical surroundings.

Biophilia is not just the love of nature but our inherent physiological and psychological need to be closely connected to it. The human brain responds to natural patterns and cues; the daily and seasonal cycles. The Biophilia hypothesis states that contact with nature and natural systems enhances healing and recovery, reduces stress, health and social problems, promotes healthy growth in children, provides restorative environments, improves cognitive functioning and attention span and promotes information processing effectiveness [71].

Bioclimatic designer, Victor Olgyay argues that the only species that are fit to survive are those which are in harmony with their environment and adapted to all internal and external forces to which they are exposed [26].

Given that most people in developed, industrialised and computerised countries (and increasingly those in developing countries) spend the majority of their lives indoors, the built environment should reflect and enhance the innate human requirement for connection to natural processes in the design, construction, operation and use of buildings, particularly in terms of natural day lighting, access to fresh air, thermal comfort, the use of non-toxic materials and non-polluting technologies which promote health and well-being and improved cognitive performance.

Thermal Comfort

Traditionally, occupants' own perceptions of whether a space in a building is providing adequate comfort, or not, was directly managed by their own personal thermal comfort experience through interaction with the building in terms of its design, construction, adaptable elements, controls (blinds, shutters, curtains, shades and windows) and management in relation to the local climate, culture and economy. Consequently, these characteristics are more likely to be well understood.

The long-term performance and impacts of the building should be the responsibility of the building designer whilst day-to-day operation should be the responsibility of occupants and must be controlled monitored and paid for by them [69]. Building design strategies will need to anticipate and accommodate new responsibilities allocated to building designers, managers and occupants [72].

Humphreys and Nicol [73] identified thermal comfort as a self-regulating adaptive system that includes both physiological and behavioural adaptations. Fundamental to the adaptive principle is if a change occurs that produces discomfort, people will tend to act to restore their comfort. Physiological adaptations to coldness include vasoconstriction, shivering and eating more food. Physiological adaptations to warmth include vasodilatation, sweating and eating less food.

Behavioural adaptations to coldness include increased activity, increased clothing, closing the posture, heating the room, finding a warmer place, closing windows, avoiding draughts, modifying the building or leaving the building. Adaptations to warmth include reduced activity, reduced clothing, adopting an open posture, separating from other people, cooling the room, finding a cooler place, opening a window, using a fan, modifying the building or leaving the building.

Psychological adaptations include expecting a range of conditions, accepting a range of sensations, enjoying a variety of sensations, accepting behavioural adaptations and accepting responsibility for control.

Daylighting

The links between lighting and health and well-being in buildings are well known. Glare, flicker, lack of contrast, inadequate illumination and unsuitable spot lighting can all add to discomfort. Many buildings are of deep plan design and therefore are unable to be illuminated by daylight to the interior. Fluorescent lighting is now commonplace and has been associated with eye strain and headaches among other symptoms. This has become less of a problem in modern lamp and luminaire technology where high frequency ballasts and reflective direction of light is better understood.

The negative effect of lack of exposure to sunlight is common knowledge and a study published as early as 1796 concluded that staying indoors for long periods causes people to appear pale, flabby, apathetic and losing vital energy [74]. Many people prefer the outdoor environment for natural light and fresh air and buildings that incorporate atria and other natural daylighting features are enjoying increasing popularity because of their impacts on the health and productivity of occupants and are an important feature of sustainable buildings. The human eye is responsible for 25% of nutrition intake, 70% of sense receptors, 90% of information we learn, processing 2 billion messages to the brain per second and contains 130 million rods and 7 million cones.

Photobiology is the science of the physiological effects of light on the body. Although most of the light energy received by the retina is relayed to the visual cortex for vision, an alternative pathway from the retina to the hypothalamus in the centre of the brain, maintains harmony within the body, and affects a cascade of hormonal responses via the endocrine and pineal glands [71]. The hypothalamus contains the biological clock, controls the nervous system and regulatory functions by sending light related information to the pineal gland which releases melatonin to the blood stream according to levels of daylight. Vitamin D is produced when skin is in direct contact with sunlight.

Enzymes and hormones regulate biological activities within the human body (e.g. digestion system). They are stimulated by selected colours of light to undergo molecular changes and their ability to function is affected by these changes. Some colours of light can stimulate certain bodily enzymes to be 500% more effective and activate or deactivate some enzymes. Therefore, using non-daylight spectrum light could affect our biological activities.

Natural daylight contains all our physiological lighting needs across the whole visible light spectrum which has therapeutic effects on humans. The natural daily cycles of dawn and dusk control our energy levels through the production of serotonin, which causes us to wake and become energised and melatonin, which induces calmness and prepares us for sleep. Deprived of lighting cues, our bodies adjust to an unnatural cycle of sleeping and waking patterns. It has been estimated that 100 bodily functions have daily rhythms per 24-h cycle.

Lack of daylight can cause depression, anxiety, fatigue and the modern diagnosis of Seasonal Affective Disorder (SAD) a disorder characterised by the lowering of mood/mood swing occurring in winter. The disorder is strongly linked to sunlight starvation which causes excessive levels of melatonin. Treatment for SAD uses full spectrum (daylight) fluorescent lamps which have shown to be effective for 80% of sufferers.

Exposure to sunlight has been found to significantly influence a range of physiological and psychological functions. For example:

- Fertility: In Finland more children are conceived during the months of June and July when the sun shines ~20 h per day, than during winter months.
- Mood: In Norway and Finland, a direct correlation has been found between decreased exposure to sunlight and a higher incidence of irritability, fatigue, illness, insomnia, depression, alcoholism and suicide.
- Exposure to sunlight produces an effect similar to physical exercise: a series of exposure to sunlight will produce decrease in resting heart rate, blood pressure, respiratory rate, blood sugar and increase in energy, strength, endurance, tolerance to stress and ability of the blood to absorb and carry oxygen.

Colour theory is also shown to have impacts on physiological and psychological responses of humans with significant effects on blood pressure, pulse rate and respiratory rate. Blue light has been found to relieve the pain of arthritis sufferers, red light relieves migraine headaches, purple light has a calming effect reducing muscle tension and yellow and orange walls improve IQ and academic achievement of school children.

Work by Ott [75] into the effects of fluorescent and natural light has shown that mice under natural daylight lived almost twice as long (16.1 months) as those under white fluorescents (8.2 months). Other research of his showed that under cool white fluorescent lamps in windowless classrooms schoolchildren demonstrated hyperactivity, fatigue, irritability and attention deficits whilst under full (daylight) spectrum fluorescent lamps in the same windowless classroom behaviour, classroom performance and overall academic achievement improved markedly in 1 month.

Other findings show that human cholesterol reduces rapidly and significantly under sunlight and therefore, reduced exposure to sunlight results in higher levels of cholesterol which is known to contribute to heart and circulatory system disease, the number one killer in many western countries. Also, levels of stress hormones were found to be higher if cool white fluorescent lamps are used instead of daylight spectrum lamps. The stress hormones are also growth inhibitors, so lack of sunlight may affect children's growth.

Supporting research by Heschong-Mahone has found that students with high levels of daylight in their classrooms progressed 20% faster in maths and 26% faster in reading in 1 year than those with the least. Similarly, students in classrooms with the largest window areas were found to progress 15% faster in maths and 23% faster in reading than those with the least. Students that had a well-designed skylight in their room, one that diffused the daylight throughout the room and which allowed teachers to control the amount of daylight entering the room, also improved 19–20% faster than those students without a skylight. Students in classrooms where windows could be opened were found to progress 7–8% faster than those with fixed windows. This occurred regardless of whether the classroom also had air conditioning [76].

Another aspect of lighting design in relation to health and well-being is the ability of the occupant to alter their exposure to meet personal comfort. Different tasks require different lighting. CIBSE guidelines recommend 500 lx for general office work and 750 lx for deep plan buildings [77]. Ideally natural day lighting should be maximised in building design incorporated with adjustable and well-designed shading devices to reduce glare and unwanted heat gain. Highly reflective tubes (light pipes) are now being incorporated into buildings to bring natural daylight to the interior which also have the additional benefit of requiring no energy to operate. Alternative strategies often rely on providing a background level of illuminance at 300 lx and offering task lighting to users so that they can control their own personal space lighting levels.

Ventilation and Indoor Air Quality

Many of the current legislative trends are resulting in increased, not decreased, emissions from buildings [78]. In the UK, current building regulations are driving designers towards the use of air-conditioning and away from less energy-intensive and arguably healthier natural ventilation of buildings [79]. According to Professor Roaf et al. [69] ventilation is poor in many closely controlled buildings and Seppanen and Fisk [80] state that relative to natural ventilation air conditioning was consistently associated with a statistically significant increase in the prevalence of one or more SBS symptoms by ~30–200%. Not least because poor maintenance practices lead to unhealthy symptoms and contaminants, related to the characteristics of heating, ventilating and air-conditioning (HVAC) systems. The Building Research Establishment [3] found that air inside buildings could be up to 10 times more polluted than the air outside.

Indoor air contains microscopic particulates—perfumes, bacteria, viruses, dust mite allergens, respiration particles, pollen, mould; combustion products—carbon monoxide, nitrogen dioxide, sulphur dioxide, hydrocarbons; and volatile organic compounds (VOCs)—benzene, formaldehyde, chlorine, synthetic fibres, PVC and many other chemical products. Pollution comes from paints, preservatives, insulation, adhesives, carpets, soft furnishings, furniture, cleaners and air fresheners. Other sources include timbers, ply, particle board which may have been treated with preservatives, glues, paints and varnishes. Fire retardant chemicals add to this potentially toxic cocktail of chemicals.

Newer construction techniques and materials result in buildings becoming increasingly airtight. In such buildings the envelope has been designed to minimise leakage of air-conditioned air to the outside but have also stopped fresh air permeating to the inside. Sealed windows and artificial air conditioning plants (heating, ventilation and air conditioning—HVAC) are thus needed to maintain warmth or coolness within. However, this trend seems to be allowing a build-up of noxious air. HVAC plants have been identified as a potent cause of SBS. Microbes and bacteria have been found in ducts and filters which are then circulated throughout the internal space. Selecting non-polluting, non-toxic and non-synthetic materials, in essence more natural and organic materials, for the building fabric, services and finishes at the outset will reduce the burden of poor indoor air quality on occupants.

Indoor air pollutants such as VOCs and particulate matter have been linked to a number of health issues including respiratory and cardiac problems. Smedje and Norback [81] investigated the impact of improving ventilation systems in schools on the incidence of allergies, asthma and respiratory problems in pupils. They found that incidences of asthma symptoms, but not allergies, were reduced in classrooms with more appropriate ventilation systems. An American study by the Environmental Protection Agency (EPA) showed that school buildings are particularly susceptible to the effects of indoor pollution because they tend to have high occupancy levels, tighter budgets and a higher number of pollutant sources. Children are particularly vulnerable due to their size and in terms of their stage of development as children breathe a greater volume of air in proportion to their body weight than adults [82].

Buckley et al. [83] suggest that poor indoor air quality is a significant problem which increases staff and pupil absenteeism in many schools, which in turn reduces productivity and performance. CO_2 levels have also been closely related to the performance of pupils and staff. A study by Coley and Greeves [84] found that in classrooms where CO_2 levels were high, pupils performed significantly less well on tests of cognitive function which indicates a reduced ability to concentrate. The researchers concluded that the size of the decrement is of similar magnitude to that observed in the course of a morning when students skip breakfast.

Ensuring adequate ventilation is the standard method for reducing the level of indoor air pollutants and CO_2 concentrations, thereby improving air quality. However, research has shown many schools, and by inference other buildings, fall below the minimum standards for ventilation.

Field surveys carried out by Kukadia et al. [85] at 8 primary schools showed direct association between the environmental conditions in classrooms and pupils' cognitive performance. The air quality conditions in the classrooms were found to be inadequate for teaching activities during about 35% of the school hours. The pupils and teachers in the classrooms studied were exposed to unacceptably poor air quality conditions with CO_2 concentrations of up to 3.5 times the existing recommended levels of 1500 parts per million (ppm).

In 16 classrooms interventions were made to improve the ventilation rate. As a result of the interventions the provision of outdoor air to the classrooms was improved from the prevailing levels of about 1 L/s per person to about 7–8 L/s per person. The results of computerised performance tasks performed by more than 220 pupils showed significantly faster and more accurate responses for Choice Reaction (by 2.2%), Colour Word Vigilance (by 2.7%), Picture Memory (by 8%) and Word Recognition (by 15%) at high ventilation rates compared to low ventilation conditions.

The study also highlighted that staff were often reluctant to open windows to increase ventilation and reduce levels of pollutants because noise levels would increase and cold air would enter the classrooms. Interestingly, the research showed that the ventilation rate could be increased by greater window use with only a small impact on thermal comfort. The research provides strong evidence that poor ventilation rates in classrooms significantly reduce pupil's attention and vigilance, and negatively affect memory and concentration.

Materials

In the late twentieth century, industry and commerce has produced around 70,000 new synthetic materials and chemicals according to Bradley [68]. Of these, <2% have been tested for safety to humans and up to 70% have not been tested at all.

Around 1000 new chemicals and materials are marketed each year, without the full cost to human or ecological well-being considered. We take in this chemical cocktail via breath, skin, water and food.

In the quest for healthy sustainable buildings the complexity facing building designers cannot be overstated. For material selection procedures to effectively review environmental and biological impact, the analysis must consider the impact from the extraction of raw materials through manufacture, use, maintenance, reuse, or disposal, in order to fully assess the global, local and personal hazards that may accrue from all the materials and components that make up the building.

This is often beyond the resources available to the design team who invariably rely on product manufacturers to supply basic data about their materials and components which, as already stated, are often untested in terms of their effect on human health throughout their lifecycle [68]. It is not unreasonable to assume that a considerable amount of time would be needed to check the environmental and biological characteristics of materials. This highlights the need for industry-recognised databases on the sustainable credentials of construction materials. Some do exist such as the BRE's Green Guide to Specification [86, 87] but are by no means adopted by the mainstream construction industry. Also, the designer and their client would reasonably expect some form of warranty from the manufacturer of an untried new product. It is to be doubted that manufacturers are in a position to guarantee that their product is free of any hazard to health.

Two branches of medical science may be involved in assessing the impacts of construction materials on human health: toxicology and epidemiology. Toxicology can provide an early indication of possible health effects by means of experimental work. Epidemiology involves observation of a sample population in order to assess some suspected environmental hazard. Work in these fields inevitably lags behind new material developments and means that building occupants could be at risk in the intervening period.

The increasingly complex mix of materials and chemicals is encapsulated within airtight buildings during the construction phase and subsequently become virtually invisible to the general building user. All materials degrade over time and can leach out or 'off-gas' to the internal environment with consequent impacts on occupant health and well-being. The more synthetic, inorganic, chemical-based and man-made materials tend to have more complex compositions, such as the wide variety of plastics available to the construction industry that give off volatile organic compounds (VOCs), and therefore potentially pose a greater risk to health than naturally occurring organic materials, but even these can pose a health hazard as many natural materials can contain harmful concentrations of radon and formaldehyde.

Rarely are the materials and chemicals that are 'locked-in' to a building released in a single moment and the demolition of buildings in carefully controlled environments, illustrates the hazardous nature of many construction materials, particularly evident with the widespread incorporation of asbestos in the post-war period. A powerful and tragic example of the toxic mix of materials and chemicals incorporated into modern buildings occurred on 11th September 2001 when a terrorist attack caused the twin towers of the World Trade Center in New York to collapse resulting in the pulverisation and co-mingling of the main constituents of the construction materials producing vast clouds of dust.

The dust contained a significant proportion of the tower's constituents, such as concrete, glass and gypsum. Photographs of 'ground zero' show piles of shattered steel and aluminium cladding but show virtually no signs of the tens of thousands of tons of concrete that constituted the 4-inch-thick floor slabs of the tower's 110 floors. Most of this had been pulverised into a fine dust. A scientific study [88] conducted a detailed analysis of selected dust samples.

The study performed inorganic and organic analysis. The inorganic analysis identified radionuclides, ions, and asbestos; and the organic analysis identified numerous dangerous types of compounds, including: polycyclic aromatic hydrocarbons (PAHs); polychlorinated biphenyls, dibenzodioxins, and dibenzo furans; Phthalate esters; and brominated diphenyl ethers and carcinogenic materials. Exposure to and inhalation of the toxic dust has resulted in serious illness and death beyond the normal physiological effects of breathing inert dust particles and offers compelling evidence of the toxicity and carcinogenic content of modern construction materials.

Few designers make use of the information available from medical science research unless it is embedded in legislative requirements in a readily accessible format and with a detailed breakdown of the content of the building materials and components under consideration. It is interesting to speculate upon who should be responsible for making information on the content of building products available. Manufacturers often claim the need for commercial confidentiality which makes open access to detailed information difficult.

There are many factors that affect users' perceptions of the buildings they occupy and their physical and psychological well-being. Much of the research undertaken to date has focused on aspects of the internal environment such as thermal comfort, lighting and indoor air quality with less emphasis on the actual materials used. A recently commissioned research project [89] focuses on end users' comfort levels and their aesthetic and emotional responses to materials selection. Looking at colour, haptic (touch and feel) properties and emissions of materials amongst other characteristics, the project explored occupants' perceptions of how they are influenced by the choice of materials used, specifically in offices, classrooms and hospitals.

Responses were gathered via focus groups, qualitative questionnaires, sensory experiences, interviews and an online survey. The building elements studied were limited to ceiling, wall and flooring materials including windows and doors. When occupants were asked to assess the impact they believe that interior materials have on their physical and psychological well-being, 88% thought that they had a 'great' or a 'very great' impact on psychological well-being compared with 64% reporting a 'great' or a 'very great' physical impact.

As awareness of the need for healthier and sustainable buildings grows it is inevitable that the detailed information upon the contents of building materials must be made available. The only decision is whether this will be by voluntary code or by means of regulation. An important factor will be the need to establish confidence and trust between manufacturers and designers. Designers must use the information wisely and manufacturers will need to assure us all that they are honestly addressing these issues. Proper comparative risk assessment of the alternatives seems to be the best option. Further research is required to improve our knowledge and understanding of the health effects from chronic exposure to a wide range of contaminants in indoor air.

Acoustics

Noise can be an environmental stressor and can have adverse effects on the well-being of occupants. Steg and Gifford [90] state that relationship between noise level and annoyance is moderated by social-psychological factors. They assert that individuals tend to be less annoyed by noise when they have a positive attitude towards the source of the noise, or, people who have control over a source of noise are much less annoyed than those who do not control it, and people who believe the noise has an important purpose are less annoyed than those who do not.

Sources of noise in buildings are air-conditioning plants, outdoor noise filtering indoors, office equipment and 'people' noise. According to Shield and Dockrell [91] the general effects of chronic noise exposure on children are deficits in sustained attention and visual attention; poorer auditory discrimination and speech perception; poorer memory for tasks that require the high processing demands of semantic material; and poorer reading ability and school performance.

Therefore, good architectural design is critical to ensure that the fabric is designed to attenuate external noise and good acoustical strategies are adopted to reduce internal noise disturbances and the building services design and associated equipment selection are carried out to ensure noise from services does not impact on health and well-being. Achieving correct solutions is not always simple given the constraints imposed by architecture, structure, services, distribution and cost—even in properly co-ordinated projects. This is covered to some extent in Part E of the UK building regulations.

Obesity

A growing body of research reveals that the design of the built environment makes an important difference in whether people are physically active. Population shifts to cities and suburban areas, with streets designed for cars instead of pedestrians or cyclists, along with building design that prioritises lift and escalator usage in preference to stairs has helped to contribute to the steadily increasing obesity rates in developed countries.

The societal and financial costs of these design decisions are dramatic. According to the Department of Health [92] healthcare costs of obesity in the UK is estimated to be £5.1 billion (due to the diseases such trends cause, from diabetes to heart disease. Currently 15 million people in the UK are classed as obese and if trends continue it is estimated by the NHS [93] that this figure could rise to 26 million by 2030 [93].

Knowledge about urban design and building design strategies is increasing and how these are linked to greater physical activity and healthier eating. A new LEED innovation credit focuses on building design strategies such as making staircases more prominent and attractive, including bike storage in buildings, and creating areas for adult and child exercise. These are changes that benefit not just people's health, but the vitality, sustainability, and economic competitiveness of communities.

The debate on comfort, health and well-being covers both the obvious and the less frequently addressed criteria which affect our physical and physiological reactions to our environment. It also extends to consideration of the psychological factors which are generally outside the professional competence of the architect and engineer. Designers may in future be guided by medically oriented research into design solutions which could alleviate the effect of factors, such as perceived lack of control over 'personal' environment, poor management of indoor environments or the selection of unsustainable construction materials.

2.6 Economic Perspectives

There is a widespread perception that the protection of the environment is contrary to economic interests. Conventional economics is concerned largely with continual economic growth—measured by rises in a country's Gross Domestic Product (GDP). Historically there has been a close correlation between economic growth and environmental degradation: as communities grow, so the environment declines. This is clearly demonstrated on graphs of human population numbers, economic growth, and environmental indicators (see Figs. 2.6 and 2.7). Economic sustainability is the ability of an economy to grow without incurring corresponding increases in environmental pressure. This is commonly termed decoupling.

In 2011 the International Resource Panel (IRP), hosted by the United Nations Environment Programme (UNEP), warned that by 2050 the human race could be using up natural resources at three times its current rate of consumption unless nations can make serious attempts at decoupling economic growth and environmental stress [94] (Fig. 2.8).

Conventional economics states that as a commodity or service becomes scarcer the price increases and this acts as a restraint that encourages frugality, technical innovation and alternative solutions. However, this only applies when the product or service falls within the market system. As the environment is generally treated as an economic externality it is un-priced and therefore overused and degraded. Market strategies can be used to internalise these externalities through tax reforms and financial incentives, carbon trading and the payment for environmental services. Societies generally recognise some ecosystems services such as food, water and energy as they pass through the market, the value of others such as climate regulation, the delivery of clean water and green spaces for recreation are much more difficult to quantify. A recent report by the UK National Ecosystem Assessment (UKNEA) [96], a collaboration between social and economic scientists, has measured the benefits that the natural environment provides to society and continuing economic prosperity and places a financial value on the natural environment. Traditionally, the common view has been that caring for the environment means extra financial burdens.

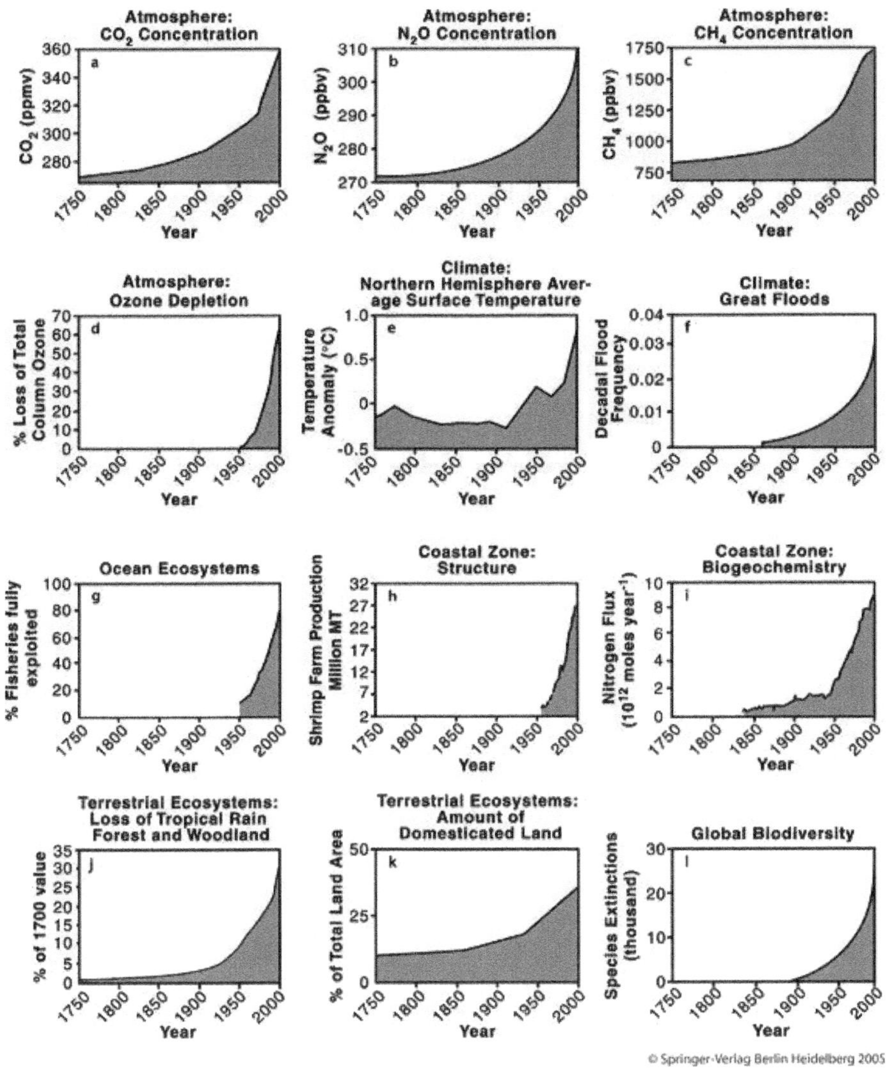

Fig. 2.7 Metrics of global change. *Source* Steffen et al. [95]

But the UK NEA shows that there are real economic reasons for safeguarding nature by providing values for a range of ecosystem 'services' the natural environment provides. For example, the amenity benefits of living close to rivers, coasts and other wetlands are worth up to £1.3 billion per year to Britain, making a strong economic rather than emotional case for their protection and preservation and the health benefits of living near green spaces are worth up to £300 per person per year. The report concluded that population growth and climate change are likely to put additional pressure on ecosystems and actions taken now will have benefits far

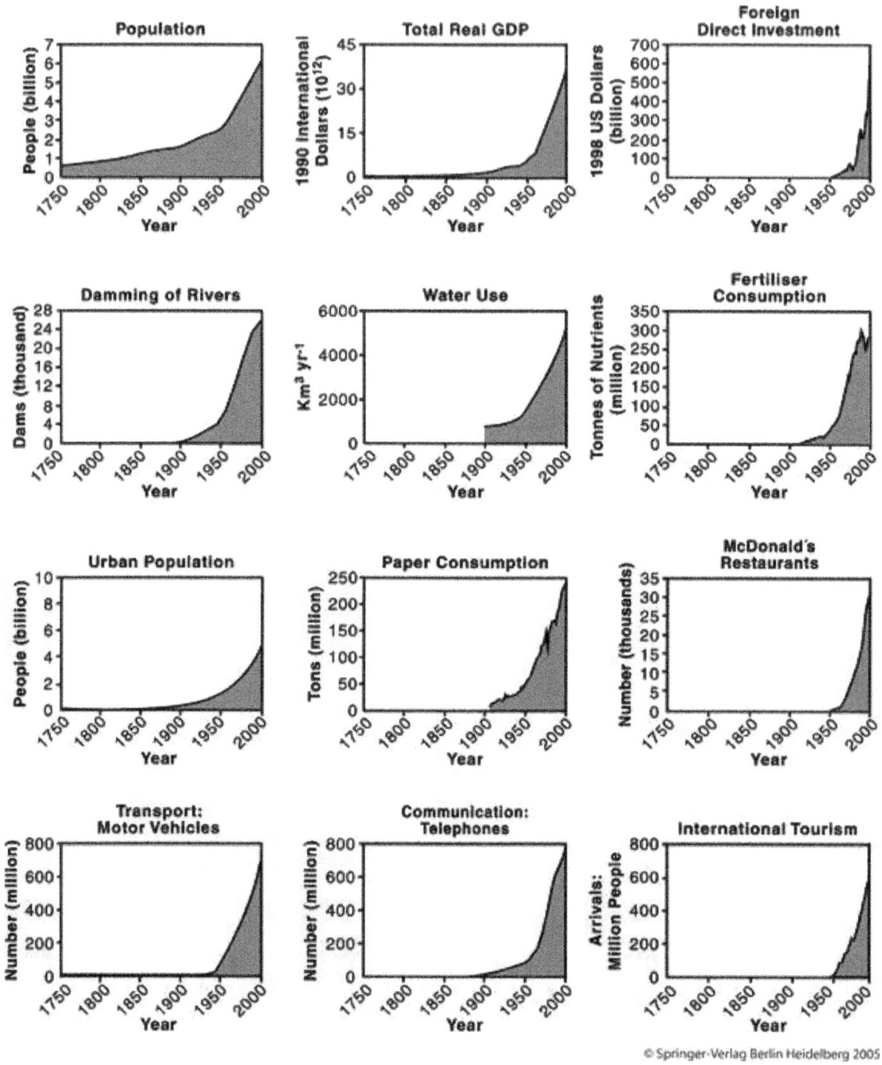

Fig. 2.8 Drivers of global change. *Source* Steffen et al. [95]

into the future. It stresses the need for a more collaborative approach to enhancing the environment including government, the private sector, voluntary groups and the public.

The Stern Review on the Economics of Climate Change [2] reports in detail the effects of global warming on the world economy. The Stern Review's main conclusion is that the benefits of strong, early action on climate change far outweigh the social, economic and environmental costs of not acting. For example, annual damage to UK buildings due to flooding from sea and rivers currently totals around £1.3 billion and is projected to rise to between £2.1 billion and £22 billion by the 2080's due to climate change [51].

According to the Review, without action, the overall costs in dealing with the consequences of climate change will be equivalent to losing at least 5% of global gross domestic product (GDP) each year and potentially 20% of GDP taking into account a wider range of risks and impacts. In essence the Review looks at cost–benefit analysis that factors in the 'opportunity cost' of not acting and the 'option value' (willingness to pay) of waiting for further information before acting.

The Review proposes that 1% of global GDP per annum is required to be invested in order to avoid the worst effects of climate change. In June 2008, Stern increased the estimate for the annual cost of achieving stabilisation in carbon emissions to 2% of GDP to account for faster than expected rates of climate change.

The Stern Review has received criticism from a wide range of sectors which include; questioning the scientific consensus on climate change [97], the impact of recommendations on UK competitiveness [98], discounting errors in calculating the possible economic damages of future climate change [99], bias in selecting references [100] and accusations of 'eco-fundamentalism' [101]. It has also received positive critical responses highlighting the need to act now, the potential for green businesses to flourish and giving momentum to tackling environmental problems.

Simply measuring economic growth through GDP has been strongly criticised. Professor Tim Jackson, professor of sustainable development at the University of Surrey [102], p. 2 states 'we're living with the legacy of our growth-obsessed society.' He describes the economy, based on growth over the last two decades as an 'illusion' and cites the decline in manufacturing, weak export markets, lack of consumer confidence and chaotic financial sector as evidence of the failure of basing prosperity on a 'meaningless indicator—gross domestic product.'

Professor Peter Victor, author of *Managing Without Growth* states:

> We have to run our economies in a way that places a lesser burden on the planet. We've overreached the capacity of the planet to support us and we need to give up the pursuit of growth as a primary policy objective.
>
> (Victor [103], p. 3).

In their book, *The Green Building Bottom Line*, Melaver and Mueller [104] propose a move towards an alternative form of capitalism. They call this 'capitalism with a difference' as opposed to 'indifferent capitalism.' This is summarised in Table 2.2.

This indicates that a significant change in economic behaviour on an individual, organisational and institutional level is required in order to achieve more sustainable business practices that can be applied to the development of sustainable buildings. By engaging in sustainable business practices, it is argued that self-actualisation takes place through experiential learning, awareness and ultimately changes in sustainable behaviour take place.

This change requires a higher ethical stance or the recognition that it is prudent business sense to move in a more sustainable direction. Adherents to the narrative of indifferent capitalism might look to government and its use of regulations, subsidies

Table 2.2 A tale of two capitalisms

Concept	Indifferent capitalism	Capitalism with a difference
Basis in economic theory	Adam Smith, The Wealth of Nations (1776): Free circulation of money, goods, labour, specialisation providing for market efficiency	Adam Smith, The Wealth of Nations (1776) plus The Theory of Moral Sentiments (1759): Balance commercial liberty with moral sympathy, holding that the efficient market is composed of small enterprises located in community
Bottom line	Financial performance only	Includes social and environmental metrics with financial performance
Governance	Shareholder theory: business is answerable to its shareholders	Stakeholder theory: business is answerable to a broader range of people beyond shareholders to employees, community
Scope	Globalisation	Internationalisation (each country sets its own rules) or localisation (self-sustaining local economies)
Notion of growth	Maximise growth: the bigger the pie the better	Optimise growth: smaller pie, more equitably distributed
Integration	Efficient integration of global supply chain	Effective integration of economic, social, and environmental issues
Mobility	Highly mobile in terms of investment, production, capital; will move quickly in and out of markets based on cheap costs and high returns	Place-centric, tending to put down long-term roots in a particular locale. Strong investment in community
Value and investment threshold	Focus on quarter-to-quarter short-term returns; value based on stock price	Greater focus on long term value, total return
Company profile	Large transnational companies, typically publicly traded, with short-term investment horizon	Often (though not necessarily) smaller companies or larger publicly traded companies with stable, long-term shareholders

Source Melaver and Mueller [104], p. 10

and incentives to change the way business is conducted, in order to address environmental and social justice issues more effectively. However, sole reliance on the current paradigm of indifferent capitalism to respond to urgent environmental issues is a strategy that is simply too slow.

A tool used by economists and market analysts, the ‘s-shaped curve’ illustrates the time it takes for an innovation to become widely diffused, hit a tipping point, and become adopted. This is slow when a critical mass of innovators and early adopters is absent. Change is accelerated as these variables increase requiring organisations to become more responsible for their social and environmental consequences of its activities. This requires a change in systemic thinking and is often driven by pressure being exerted from outside the organisation by bodies working for social and environmental justice. This is highlighted by the widespread adoption of Corporate Social Responsibility (CSR) as a driver for organisational change.

Sustainable business practices integrate ecological concerns with social and economic ones which can result in triple bottom line benefits, e.g. investment in energy efficiency can increase profits by reducing costs potentially leading to job creation and increased market share due to an improved public image. The idea of sustainability as a business opportunity has led to the formation of organisations such as the Sustainability Consortium of the Society for Organisational Learning, the Sustainable Business Institute, and the World Council for Sustainable Development.

Using whole-life analysis avoids decisions being made based solely on the short-term costs of design and construction. Often the longer-term maintenance and operation costs can be a significant proportion of the whole-life cost, and for some projects will far outweigh the initial design and construction costs. Whole-life thinking is an essential component of sustainable design.

Economic Benefits of Sustainable Buildings

A recent study by the World Business Council for Sustainable Development [105] found that the building industry greatly overestimates the cost of sustainable building while underestimating the amount of greenhouse gases emitted from the construction and use of buildings. According to the study, respondents estimated that implementing sustainable building practices would increase costs by 17%, when the premium is in fact closer to 5%. Building-related greenhouse gas emissions were estimated at 19% of the global total, when the figure is actually closer to 40%.

UK and US studies, referred to below, show that there are real and quantifiable benefits of sustainable buildings in terms of; capital cost savings, reduced running costs, increased investment returns, increased productivity, staff recruitment and retention, more efficient resource use and major image and marketing advantages.

Decisions made early on in a project to incorporate sustainable features are likely to result in less of an increase in capital costs than those made at a late stage and may result in significant savings over the life of the building. Two Building Research Establishment studies [106, 107] investigating the costs of various building typologies show that to achieve a 'Very Good' rating would require an increase in capital costs of between −0.4 and 1.3%. An 'Excellent' rating would require an increase in capital costs of between 1.9 and 4.1%.

To scale-up green buildings to become the norm requires investors to perceive sustainable buildings as financially attractive. Common and comparable certification standards can assist potential investors and developers in assessing the merits of green building projects. There are perceived risks in that sustainable buildings can be seen as a short-term trend with a narrow marketplace appeal that runs counter to longer-term investment strategies. One of the problems is that a disconnect seems to exist between those who pay the upfront costs of the development and those who ultimately benefit from the savings in the longer term. To correct this 'advanced incremental funding' has been developed, which involves a financial partner issuing a loan to the property owner, who then repays the loan over a predetermined period, allowing the cost of the loan to be passed on to tenants through common element charges. After the loan period, the energy efficiency savings then accrue to owners and occupiers.

As tenants provide the ongoing demand for buildings the benefits of occupying a sustainable building must be understood and there is evidence that sustainable buildings achieve higher rents and therefore reduce investment risks.

In a research report, evidence revealed that effective rents for certified green buildings are about 6% higher than rents for comparable buildings located nearby [108]. Achieving high efficiency ratings also acts as a safeguard to minimise the effects of future energy price increases, recognising carbon emissions as a real cost, with future cost and taxation implications and therefore it makes strong economic sense to develop greener buildings. Investors are also increasingly seeing socially and environmentally responsible investment as a priority moving towards longer-term investments in communities and maximising returns by optimising life-cycle capital investment. Institutional investors such as pension funds have a longer-term outlook and offer better alignment with sustainable projects.

Commonly understood economic principles of economies of scale suggest that institutional capital wants to commit to multiple projects over an extended period of time. However, replication does not generally lend itself to sustainable construction projects which tend to focus more on sensitivity of place, people and community, factors which can be perceived as adding another layer of complexity to the standard economic model.

Tax credits focussed on new markets include social as well as financial components. A company with a triple bottom line orientation, focussed on social and environmental consequences of its activities is more likely to be developing in urban core areas where development can be a catalyst for revitalising a community. Tax credits can make a project significantly more financially viable. A triple bottom line focussed development has to deliver on the financial metric first and foremost.

According to a report published by the Department for Communities and Local Government [109] it was found that the average cost of building a home to code level 3 of the Code for Sustainable Homes (CSH) has fallen by more than 70% over the past three years. It was found that the average additional cost has fallen from £4458 in 2008 to £1128 in 2010. The report concluded that this was largely due to operational efficiencies and economies of scale in the higher volume of procurement and supply chains for sustainable technologies and that decreases in costs were attributable to market pressures rather than greater experience in building to CSH.

The report highlighted the ability of builders to develop optimal solutions which included focussing on the fabric first then to focus on renewable technology which yielded greater efficiencies in construction through standardisation. Even though the cost of renewable technologies was observed to be decreasing due to a large volume of supply, a significant skills gap was identified for installing specialised technologies resulting in high labour rates but the skills base is likely to expand to meet future demand.

In addition, the report looked at overall price trends across the building industry and observed that the building industry is a complex area given the significant level of uncertainty in both technical solutions and market changes which is exacerbated when projecting costs forwards for six years or so in the context of the currently 'depressed' market.

Alternative materials exist which can deliver higher performance qualities more efficiently, but because they are not mainstream, they will tend to be more costly. As the industry adapts to lower U-values we are likely to see material costs fall through efficiency gains in bulk procurement of these new materials.

The market for renewable technologies is relatively mature on a European and global level, UK supply chains are still growing to meet demand. Advances are still being made that should reduce the unit cost and it is expected that the cost of renewable will decrease through volume procurement and technology advances.

Operational costs are diverse ranging from utilities such as energy, water and waste, staff costs, rents, maintenance, management costs and transport and computing. Yates [110] found that operational resource use costs account for ~5 times constructional costs over the typical 60-year life of a building and typical staff costs account for ~200 times the construction costs over the same period.

Sustainable buildings in their efficient use of resources have the potential to make significant cost savings as the price of energy increases and investment in alternative forms of energy offset the demand for ever-more redundant finite fossil fuel-based systems. A change from air-conditioning to naturally ventilated space results in significant operating savings to the value of about £100–150/m^2 of gross floor area [110].

Energy efficiency is now a significant market driver with the introduction of mandatory Energy Performance Certificates (EPCs) Display Energy Certificates (DECs) improved Building Regulations (Part L) and the Carbon Reduction Commitment (CRC) for large public and private sector organisations which are responsible for around 10% of the UKs carbon emissions [111]. With the introduction of the Climate Change Levy, building occupiers have been given a strong incentive to operate their buildings more efficiently. The levy can increase energy bills by about 10%, leading major property owners to carefully consider the options for reducing consumption of energy. The provision of sustainable and healthier indoor environments has positive cost implications through the well-being of staff, resulting in higher retention rates, less illness, absenteeism and increased productivity.

Leaman and Bordass [112] in their extensive workplace studies show that there is a strong and significant relationship between perceived comfort and perceived productivity. Workplace ecology connects the occupier with their environment through psychological (perceptual) responses and physiological (biological) responses. A detailed workplace study, the Gensler Office Survey by Clements-Croome [113] applied these two factors to top-performing companies, applying the Work Productivity Index (WPI). Under the four variables of focus, collaboration, learning and socialising the top-performing companies did a better job of supporting their employees' work modes by 12% more than average-performing companies.

The survey found that office workers claimed an improved workplace would increase employee productivity by 19%; equivalent to a £135 billion annual increase in the UK's service sector output. 79% of office workers consider quality of working environment very important to job satisfaction and over a third stated that the working environment is a factor in accepting or rejecting a job offer; 58% believe their office has not been designed to help them do their job. The survey concluded that office workers want a better working environment and identified personal space (39%), climate control (24%) and daylight (21%) as crucial factors for a good working environment.

In a separate study by Romm [114] of five companies it was found that investment in daylight design improved productivity by between 5 and 14% leading to energy savings of between 22 and 64% by offsetting the need for artificial lighting with typical payback periods of between one and three years saving one company $300,000 per annum. Personal climate control was shown to have increased worker productivity by 7% and reduced energy demand by 40%.

Leaman and Bordass [112] recognise that it is difficult to link human productivity to space conditions but questionnaire data from their extensive Post-Occupancy Review of Building Engineering (PROBE) studies show that losses or gains of up to 15% of turnover in a typical office organisation might be attributable to the design, management and use of the indoor environment and their data shows that characteristics which improve perceptions of individual welfare contribute to better energy efficiency. They also highlight that the majority of occupied buildings do not exhibit self-reinforcing qualities (lack of user control) and many are unmanageably complex.

Leaman and Bordass state 'the perversity of human behaviour means that building occupants can report that their environment is unacceptable despite it falling within industry-standard comfort guidelines' [112], p. 19. From results based on the PROBE studies the four key variables that have a critical influence on the behaviour of a building as a system have been highlighted as; personal control, responsiveness, building depth and work groups. The PROBE studies are discussed in greater depth in Chap. 4.

Peoples' perception of control over their environment affects their comfort and satisfaction and people are more tolerant of conditions as the perception of personal control—windows, switches and blinds etc.—increases. Even being able to make small changes to their immediate environment, occupants can make the difference between intolerable conditions to tolerable ones without management intervention.

Buildings which have been found to work well and achieve high user satisfaction scores have the capability to meet peoples' needs very rapidly. This can be through personal control or facilities management response to complaints and can go beyond the common parameters of temperature, lighting and ventilation control to the ability to reconfigure spaces. Poor office layouts have been estimated to lead to a decrease in productivity of 10–20% [16].

Post-occupancy research by Leaman and Bordass [112] suggests that overall satisfaction and productivity will decrease with increasing depth of the building. Buildings become disproportionately more complicated as they get bigger and deeper. Deep forms tend to use artificial lighting and air conditioning more and impact on spatial and behavioural group dynamics with higher staff numbers and higher dependence on technology and management whereas shallower plan forms tend to cost more and are less space efficient but lend themselves to cheaper, more domestic envelope construction, cheaper services and lower management costs. It was found that a depth of 12 m across the building is optimal for human performance variables.

Leaman and Bordass [112] also found that air-conditioned buildings produce poorer perceived productivity scores than naturally ventilated buildings. It was found that occupants tend to prefer natural ventilation as the default state for winter, spring and autumn, and air conditioning during hot and humid summers. Clearly, as summer temperatures are predicted to rise due to climate change the need for cooling becomes an issue, which can be solved through passive ventilation strategies rather than relying on energy-intensive air-conditioning systems. There is also strong evidence that air-conditioning systems are less healthy for occupants as within their mechanisms they can harbour pathogens and they do not always circulate the required amounts of fresh air.

The research also showed that generally, perceptions of productivity are higher in smaller and more integrated work groups and there is a preference for cellular offices for individuals or small work groups which have greater perceived user control for temperature, lighting and ventilation—perceived control declining with work groups bigger than five people.

To summarise this section, there is a learning curve associated with designing, constructing and operating sustainable buildings. For both public and private owners and developers of green buildings, subsequent green buildings generally cost less than the first. The trend of declining costs associated with increased experience of sustainable construction is evidenced by significant research findings in a report by the U.S. Green Building Council of 30 sustainable buildings [115]. Three reported LEED Silver buildings in one district incurred cost premiums of 2, 1 and 0% respectively whilst in another area saw the cost premium of LEED Silver buildings drop from 3–4 to 1–2%.

The same report found that the financial benefits of greening buildings are more than 20 times as high as the cost of going green. The report concluded that:

> ...the analysis of costs and benefits of sustainable buildings and use of conservative and prudent financial assumptions provides a clear and compelling case that choosing to design sustainable buildings is extremely cost-effective, and represents a fiscally far better design choice. Building green is more fiscally prudent and lower risk than continuing to build unhealthy, inefficient buildings. [115], p. 11

2.7 Technological Perspectives

From the beginnings of the evolution of the human species people have used their natural and innate ingenuity to innovate and use materials and new techniques to provide shelter and adapt to their surroundings, from applying mud to a lattice of sticks, wrapping an animal skin around a wooden frame or shaping blocks of ice, stacked so as to form a dome. These were highly natural structures that, by modern definitions, would qualify as highly sustainable.

Technology has evolved and adapted over time but with essentially the same fundamental techniques and materials until the modern era when highly processed materials and increasingly sophisticated practices were applied to buildings, typified by the development of steel, concrete and glass skyscrapers requiring energy-intensive technologies such as air-conditioning. Openable windows at such heights were not feasible due to the extreme air currents induced by the stack-effect with secondary health effects on occupants due to poor indoor air quality.

These technologies pushed the boundaries of design beyond the human-scale and in hindsight exhibited high levels of unintended un-sustainable characteristics; the over-reliance on air-conditioning, lack of natural day lighting, non-breathable materials, over-heating, high levels of heat loss, excessive solar gain, use of monolithic, non-recyclable high embodied energy materials, particularly plastics, limited adaptability and lack of biodiversity, resulting in what is now commonly known as sick building syndrome, a recognised medical condition. The buildings often required high levels of costly unplanned and responsive maintenance and repair.

They were often built with the best intentions in the name of progress and innovation but in mainstream circumstances the adoption of the technology was driven by low cost and low-quality industrialised methods of design and production, as illustrated by the tenement blocks of the mid-late twentieth century.

The impact on the behaviour and practices of architects, engineers and constructors was to develop more mechanistic and technical solutions to accommodate growing urban populations and the requirement of specialist operators to control the indoor environments, removing user responsibility to operate their own conditions, according to their perceived physiological and psychological needs within relatively narrow parameters. The impact of this along with the scale of the developments had a de-humanising effect and at the very least provided a setting for anti-social behaviour and at worst created it.

The 'New Modernist' movement claimed that tall or dense buildings were more sustainable and carbon beneficial than traditional low-rise construction. However, ecological architects and climate change experts have begun to brand glazed skyscrapers as 'profligate dinosaurs.' Alan Short, Cambridge Professor of Architecture argues that in the future buildings that use large amounts of energy, and big glass office buildings, will be pariah buildings and that people won't want to rent them [116]. Therefore, there is a growing re-assertion of a more common-sense movement, identifying sustainability with modest architectural scale, as also happened in the reaction against system-built high towers.

These issues are now generally understood and the adoption of more human-centred sustainable technologies is becoming increasingly commonplace, recognised by the mainstream construction industry bodies such as RIBA, RICS and BRE evidenced by a myriad of publications on the over-arching topic of sustainability. Sustainability now impacts upon every aspect of the building process and with the increasing demand for sustainable buildings, technical solutions are inevitably sought to reduce the environmental impact of the built environment.

A lot is now being made of 'smart' or 'intelligent' technologies. According to Dexter:

> The term 'intelligent building' is used to designate one that can adapt its mode of operation to the needs of occupants and the current operating conditions, and also learn how to optimise its performance. Such a building would also be able to detect and compensate for the presence of faults, thereby ensuring the maintenance of desired comfort conditions in all circumstances, while minimising energy consumption, environmental pollution and maintenance costs at all times.
>
> (Dexter [117], p.6)

Smart materials are seen as materials that can respond to the environment and provide some form of feedback or control on the environment such as smart windows with optically switchable electro-chromic glazing or smart walls with a variable U-value. This is achieved by incorporating microencapsulated phase change materials (PCMs). The PCMs store large amounts of heat energy when the ambient temperature rises and release it when the temperature drops, it is claimed improving comfort and reducing the energy demand of the building [118].

Technological innovations can undoubtedly increase resource efficiency to reduce the environmental impact of building design, construction and use, particularly through advances in computer technology, such as computer aided design (CAD), building information modelling (BIM) and building energy management systems (BEMS) Sustainable innovations implicitly affect the performance of the building and arguably impact on the design, construction and management teams if utilising new and innovative technologies, methods and materials. However, they don't necessarily encourage or facilitate the pro-environmental behaviour of the building operator or user in the absence of education and awareness.

Traditionally, buildings have always required users and operators to interact with the systems designed into a building to control a number of variables to achieve satisfaction with the indoor environment.

'Intelligent' or 'smart' buildings use technology that automatically respond to changes in occupation levels, temperature, lighting etc. but go only part way in providing sustainable built environments and it can be argued that that they detract from our ability to understand, interact and engage with our built environments and limit, discourage or even undermine opportunities to change ecological attitudes and sustainable behaviours because of these design interventions. Smart buildings are not necessarily synonymous with sustainable buildings and indeed sustainable behaviour.

According to Cole 'technological innovation has led to a shifting of design responsibility in comfort provision from architects to mechanical engineering consultants and control responsibility from occupants to technology' [119], p. 10.

Clearly, building automation can result in substantial energy savings and the use of the latest control systems have found to reduce energy consumption by 50% [120] albeit under highly controlled behavioural conditions that may not always result in high user satisfaction or comfort levels.

In terms of encouraging pro-environmental behavioural change an analogy can be drawn with another environmental concern, littering. Street cleaners pick up litter because many people feel it is acceptable to litter, knowing it will be picked up, rather than engaging in the pro-environmental behaviour of not littering. The result is the same, cleaner streets, but in the former example behaviour has not changed. As with buildings, if people don't feel responsible or are disempowered or disconnected from their indoor environment because the building automatically responds to their needs their behaviour will not change and this will result in unsustainable behaviours being adopted in other, less 'smart' or 'intelligent buildings.'

It is reasonable to assume that decisions to invest in sustainable technologies are based on good financial returns. Evidence from Owen [121] suggests that people fail to take action with very short payback periods, such as insulation, whereas there can be a sudden rush to install a technology that has a long payback period, perhaps because it is a visible statement of sustainability or the technology is promoted well, such as the supply and installation of micro wind turbines that proved to underperform, particularly in urban environments.

Expected outcomes are hard to envisage. Installing insulation where the occupiers cannot afford their heating bills can increase the energy used. Heating bills are reduced so, understandably the thermostat is often turned up and the benefit is taken in warmth rather than reduced cost. This is known as the 'rebound effect' first described by William Stanley Jevons. He states that 'it is a confusion of ideas to suppose that the economical use of fuel is equivalent to diminished consumption. The very contrary is the truth' [122], p. 3. Another example of this 'effect' is the installation of reduced water flow shower heads that may well be run for longer because of perceived savings but water consumption and costs may actually increase as a result of this sustainable intervention.

In a paper by the Technology Strategy Board [123] it was reported that occupants behave in more complex ways than designers account for by opening windows, leaving doors open, generating body heat, keeping tropical fish tanks and installing plasma T.V. screens and Bell [124] suggests that architects need to develop their professional expertise to improve buildings and seek ways of integrating user involvement in building performance.

Building energy management systems (BEMS) are a form of intelligent technology incorporated into buildings. The modern BEMS is a powerful distributed microcomputer system consisting of networked outstations each with significant computational abilities. The use of sophisticated graphical user interfaces is already commonplace. More recently, techniques which mimic human intelligence have been applied to various aspects of building systems control. These have been termed 'sentient buildings' and as well as responding to physiological feedback they can also respond to psychological stimuli such as emotions and moods of occupants. Another technological advance at a relatively immature stage of development is the use of smart phones and digital technology to remotely control a building's operational systems with the potential for cost and carbon savings.

Findings from the case study analyses, discussed in Chapter 6, suggest that existing BEMS are often operated at suboptimal levels because of lack of sufficient product development, inadequate system design and poor installation or most commonly because building managers do not know how to use them, that the technology is not trusted because of past failures and as a result of unintended and unanticipated behaviours of occupants. At optimal operating levels BEMS systems have the potential to encourage sustainable behaviour by allowing the operator to monitor, target and analyse building performance and to feedback energy saving to the organisation and individuals. BEMS can also be used as a teaching and learning resource for occupants raising awareness of the environmental parameters of the building and their influence on environmental performance. This could be part of an initial induction at the commencement of occupation.

Integrating ever-increasing levels of artificial intelligence into buildings claim to lower their environmental impact in a passive way, but the long-term functionality and adaptability of these systems is relatively untested and adds a level of complexity and sophistication that may not be required if the original design concept uses more traditional and proven technologies. All sustainable buildings require good design, based on sound architectural principles incorporating basic features such as appropriate solar access, insulation and appropriate choice of materials. Arguably, all buildings would have to display the same level of intelligence for users to understand their operation, as transferring from an automated system to a manual one would inevitably cause confusion and the undermining of one or even both of the systems. Leaman and Bordass [112] suggest that designers are 'seduced' by the promise of technology, rather than by its delivered performance and tend to assign more functions to automatic control than are usually warranted. Knowingly or not, they also make the interfaces obscure.

There is a danger of an over-reliance from the rapid adoption of relatively untested technologies, a 'blind-faith' in the ability of science and technology to solve human problems. This is evidenced in other sectors of society such as the recent failure of computerised banking systems, wholesale failure of mobile phone networks and unconsidered social impacts such as the role that social networking played in the rapid spread of the 2011 riots in the UK. There are also unanswered questions about the effects of electromagnetic radiation from increasing wireless networks on the health of users and the wider population.

Cohen et al. states that:

> notwithstanding all the implications of supposedly advanced automation, our experience is that the best intelligence in most buildings lies in the occupants themselves... the challenge for designers and manufacturers is then to support them with appropriate and understandable systems with readily usable control interfaces, which give relevant and immediate feedback on (building) performance.
>
> (Cohen et al. [125], p. 57)

In other words, buildings are not inherently *intelligent* but can support intelligent patterns of behaviour. In classic management speak, 'if it cannot be seen it cannot be managed.'

Bell [124] asserts that changing behaviour is the only way to deliver real improvements in sustainability, as users already have the choice of sustainable technology, but are not taking it up as rapidly as other forms of technology. This is evidenced by the utilisation of smart metres that go some way to raising our awareness of energy consumption, CO_2 levels and energy costs and subsequent behaviour change, but a recent study has shown that our awareness of these technologies is low and there is a degree of scepticism about their adoption and that they are effective for a short period of time and then ignored [126].

The application of behavioural psychology to pro-environmental behaviour in buildings asserts that technology can encourage sustainable behaviour through feedback, interventions, information provision and education but also that human behaviour is often counter-intuitive, irrational and unpredictable when faced with purely technical solutions.

Janda [12] suggests that social expectations and consumption patterns of building users can defeat the most careful and intelligent design. Kroner [127] recognises that technology should add value to a building, and this cannot be achieved by adding even the best technology to a poorly designed, constructed, operated and used building. Therefore, 'intelligent' and 'smart' solutions based on the predictability of behaviour may operate sub-optimally.

2.8 Chapter Summary

The global challenge for the built environment sector in an increasingly globalised and homogenous world is for built environment professionals to play a key role in helping to manage climate change. The aim should be to reduce the consumption of energy and prepare structures for much altered future climates. Architects, engineers and building managers are active players in shaping the future through the design and operational choices they make. It is ethically indefensible for them to ignore the challenge of global warming.

Economic barriers can stifle the opportunity to develop new practices and technologies and limit the freedom to choose energy adaptation or innovative green technologies for short-term economic reasons. However, the global economic downturn does provide individuals, organisations, institutions and nations the opportunity to take stock of alternative visions of the future.

As well as global, national and regional environmental benefits to a more sustainable built environment, there are social advantages in terms of healthier buildings reducing the risk of sick building syndrome, increasing health and well-being and optimising the productivity of occupants plus economic benefits by investing now in a more sustainable future to offset the much greater costs of dealing with the potentially catastrophic effects of a warming planet in the future.

Incorporating ever more complex and sophisticated technological features can undoubtedly encourage sustainable behaviour but the over-reliance on scientific and technological solutions can suppress user behaviour by undermining interaction with their own built and the wider environment.

Having made the general case for the economic, social and environmental need for a sustainable built environment on a global and national level, the following chapter focuses on specific barriers, drivers and opportunities for the widespread and mainstream adoption of sustainable construction strategies from a psychological perspective based on theoretical principles and practical applications.

The following chapter also investigates how buildings and pedagogy relate to each other, with opportunities for transformational educational change, considering buildings as teaching and learning resources—buildings as pedagogy.

Chapter 3
Linking Human Behaviour, Sustainability and Buildings

3.1 Introduction

This chapter examines the dynamic and complex systems and processes by which environmentally responsible action emerges from the interaction and interrelationship of people (as individuals, organisations and institutions) and the built environment. How synthesis through collaboration between educationalists, designers, developers, operators and users can themselves change their ecological attitudes and behaviours whilst optimising and encouraging the pro-environmental behaviours of others through the built environment.

The chapter begins by establishing a definition of sustainable behaviour, how various theoretical principles relate to encouraging behavioural change in general, including non-environmental change practices and sustainable product design, and then proceeds to focus on buildings as agents for encouraging sustainable behavioural change through the application of psychological and pedagogical initiatives. The chapter concludes by presenting research findings to highlight barriers and drivers from a psychological perspective that impact on sustainable behaviour in the built environment in the context of individual, organisational and institutional practices.

By investigating relatable theoretical principles in the field of behavioural psychology and applying these to the provision of buildings that encourage sustainable behaviour through their design, construction, operation and use, it is hoped that a greater understanding will emerge of how the gap between intended and actual sustainable actions or behaviours can be narrowed.

The theoretical principles investigated include behavioural change theory, environment-behaviour (E-B) theory, environmental psychology, 'nudge' theory and actor-network theory. Looking at applied research which redefines policy and industry problems from a social practice perspective identifies new avenues for social and environmental change. This helps explore the role of new and re-visioned technologies and materials in mediating and jointly shaping everyday practice, drawing on behavioural change theory to inform the design of built environments.

J. L. Clarke, *Sustainable Buildings for Sustainable Behaviour: Theory and Practice*, Green Energy and Technology, https://doi.org/10.1007/978-3-031-98905-6_3

Lee [128] recognises sustainability as a growing area of research in the field of environmental psychology. Winter and Koger [129] focus on the relationship between psychology and sustainability reviewing various theoretical frameworks they identify as contributing to our understanding of sustainability; namely (i) Freudian/psychoanalytical, where defence mechanisms and environmental problems are linked to a changing sense of selves in relation to nature (ii) Social, where attitudes and behaviours are influenced by norms of appropriate behaviour and how we are influenced by our reference groups (iii) Behavioural, where behaviour and environment have a reciprocal relationship and research shows that behavioural change will occur where it is easy or convenient to do so (iv) Cognitive, where inappropriate or unsustainable behaviour can arise from distorted or missing information about the consequences of our actions (v) Holistic (Gestalt and Eco psychology) where conscious efforts are made to modify lifestyles in order to reduce environmental impact because of a sense of an intimate relationship with others and the ecosystem.

We often hold certain values about what our behaviours should be, that in reality do not result in action, known as the 'value-action gap' or 'intention-behaviour' gap. This is particularly true with behaviour related to sustainability which is often seen as unattainable in our working practices and everyday lifestyles, falling outside our economic and social norms. By applying behavioural science theory to the design, construction, operation and use of buildings it is hoped to understand how buildings can influence human behaviour to be more sustainable.

Shifting the emphasis from the individual and their own consumption to people as social beings acting as members of the wider society, we are able to better account for the roles of technologies, systems of resource provision, organisational issues and institutional rules and regulations.

The chapter illustrates how the *value-action gap* can be closed by studying these theoretical principles. These findings are then applied to real life examples through case study analysis involving interviews, observation and questionnaires aimed at the development of exemplar sustainable buildings used for sustainable educational purposes presented in Chaps. 6 and 7.

Research findings highlight how best practice can be achieved and adopted as the future norm. This can be applied throughout the design, construction, use and operation of buildings by enabling stakeholders throughout the entire lifecycle of the building to make sustainable choices the standard and preferred method, by encouraging economic, social and environmental sustainable thinking as the default mind-set.

3.2 Behavioural Change

Defining Sustainable Behaviour

Environmental behaviour has been defined as 'behaviour that changes the availability of materials or energy from the environment or alters the structure and dynamics of ecosystems or the biosphere' [130], p. 12. Sustainable behaviour is used as a term to describe behaviour that has a positive influence on the local, national and global ecosystems or biosphere.

A broad definition of sustainable behaviour is:

> behaviour that results in the satisfaction of our needs today without diminishing the prospects of future generations to do the same
>
> (Brundtland [34], p. 40).

Pro-environmental behaviour is a more specific term as it focuses on purely environmental outcomes of behaviour which may not necessarily result in economic or social benefits.

This can be refined by applying the triple bottom line concept of sustainable development resulting in the definition of sustainable behaviour for the purposes of this study as being the behaviour of individuals, organisations and institutions that achieve positive economic, social and environmental outcomes.

Sustainable behavioural change is defined as:

> A product of transforming socially and materially shared practices such as heating or cooling a building. Understanding how these practices are established as normal ways of living, working or playing, how and why they are changing and what opportunities exist for attempting to reorient them in more sustainable directions.
>
> (Beyond Behaviour Change [131], p. 2).

Lessons from Non-environmental Behaviour Change

Some of the greatest advances in the application of behavioural change theory have been made in areas seemingly unrelated to pro-environmental behaviour and wider environmental issues related to the built environment, such as public health campaigns. Abroms and Maibach suggest that it is important to apply lessons from public health campaigns to climate change, stating 'the public health community has much to offer. Most notably this community has both breadth and depth of experience in understanding and responding to population behaviour challenges' (Abroms and Maibach [132], p. 220).

The 'information deficit' models of public communication, that assumes people simply need more information to behave sustainably, have long been supplanted by a focus on 'ecological models,' which concentrate on two key determinants of behaviour: people and places. Corner states that 'everybody knows that people are important in behaviour change campaigns. But place factors are important too. This means influencing the decisions and behaviour of the people who control the attributes of place, such as the availability of services or products. These are the barriers that must be removed before people can implement their pro-environmental intentions' [133], p. 2.

Research shows that people are reluctant to feel vulnerable to health risks (this is only likely to be exaggerated in perceived vulnerability to the effects of environmental problems, e.g. climate change) such that exacerbating the 'fear factor' is of little use—what matters is whether people fear they are personally at risk [134]. People must receive solutions that will directly address this risk for interventions to be effective. Another crucial factor is perceived control. Any information or prompts must ensure that they perceive themselves as capable of changing—otherwise helplessness will ensue.

Abroms and Maibach conclude that:

> Multiple, overlapping campaigns that target both people and place based drivers of behaviour, including public policy, when sustained over longer periods of time, have the potential to multiply minimal effects into broad based shifts in societal beliefs, norms and practices.
>
> (Abroms and Maibach [132], p. 234)

The main lessons to be drawn from public health campaigns are that efforts must be co-ordinated across individual, social and structural behaviours [135]. It is clear then that valuable lessons can be learned for the construction industry from public health campaigns focussed on behaviour change. Parallels can be drawn in that it focuses on people and places and the need for a shift in social norms and practices through a combination of legislation, enforcement, education, individual, organisational and institutional structural change.

Lessons from Sustainable Product Design

In a recent lecture Bhamra [136] described design for sustainable behaviour as 'exploring how design (in the broadest sense) can influence user behaviour to reduce the social and environmental impacts of products during use.' This is supported by research presented in a paper by Bhamra et al. [137] undertaken at Loughborough University Design School investigating the application of Design for Sustainable Behaviour (DfSB) in two product case studies.

Strong comparisons can be drawn between sustainable product design and sustainable architectural design, buildings in essence being a product, and valuable lessons can be learned from how the field of sustainable product design has tackled the issue of how to encourage greater sustainable behaviour. Traditionally, both have strongly focussed on the supply side; how can environmental impacts be lowered through design-for-disassembly, recyclability, use of sustainable materials and dematerialisation [138].

In *Design for Sustainable Change* [139] it is asserted that Designing for Sustainability (DfS) is not just about the design of products and services, it is also about how we use those products and services and our patterns of behaviour. Various approaches to DfS are offered such as making sustainable actions easy, making people aware of their current unsustainable behaviour and presenting them with an incentive to change and, rather than own products that we require infrequently, we use them through shared ownership.

User-centred sustainable design solutions look at how users interact with a product which can strongly influence the environmental impact of that product, the demand side or human side of product design. Product designers, just as architects, can try to influence this behaviour through the products or buildings they design, such as incorporating 'eco-feedback' such as energy monitors, 'scripting' by guiding the behaviour of the user so that unsustainable behaviour is made difficult or impossible, while sustainable behaviour is made easier and 'automation' of certain functions which by-pass the user.

For many products, as for buildings, it is increasingly acknowledged that the major part of the environmental impact is caused during the operation and use phase, in particular through energy consumption, as shown from various studies by Brezet and Van Hemel [140], Stevels and Griese [141] and Abele et al. [142].

Akin to architectural design, product design has a strong tradition in the technical disciplines but research that addresses the human side of design, and how this can contribute to sustainable behaviour, is very limited. Wever et al. [138] cite one likely reason to be the customary lack of cross-fertilisation between the technical disciplines (the '*hard sciences*'), and human-focussed social science based disciplines such as user-centred design and interactive design (the '*soft sciences*') Therefore, design strategies that focus on user-centred energy efficiency and the sustainable use of products and buildings are likely to be more effective when they complement technological solutions.

In their most simplistic form, buildings may be viewed as a single product made from an amalgamation of materials, technologies, products and systems that are utilised by architects, engineers, constructors, operators and users. By applying user-centred, demand side principles to buildings, allied to supply side strategies, a more holistic approach is envisaged that can encourage greater sustainable behaviour throughout the design, construction, operation and use phases of a building's lifecycle.

Bhamra describes a number of design interventions and strategies covering three levels of intervention; (i) guiding change, (ii) facilitating or maintaining change and (iii) ensuring change. The following list shows the identified seven design interventions that can be adopted to encourage sustainable behaviour:

- Eco-information
- Eco-choice
- Eco-feedback
- Eco-spur
- Eco-steer
- Eco-technology
- Intelligent design [136]

'Eco-information' refers to design-oriented education that enables the visualisation and experiencing of resource use in order to make that resource use visible, understandable and accessible to inspire consumers' to reflect on their resource use behaviour (see Fig. 3.1).

'Eco-choice' refers to our ability to select products, materials and technologies based on their environmental impacts. 'Eco-feedback' offers links to environmentally responsible action to inform users clearly what they are doing and to facilitate users to undertake environmentally friendly behaviour through offering real-time feedbacks

Fig. 3.1 Power aware cord—seeing personal energy consumption. *Source* Bhamra [136]

providing tangible aural, visual, or tactile signs as reminders to inform the situation enabling the user to act. Figure 3.2 displays real-time and historical data for household CO_2 emissions, kWh used and costs related to electricity use.

This helps homeowners to understand the connections between their lifestyle choices and the energy use that results from them. They can then begin to see their monthly energy bills as something they can manage rather than merely accept.

Fig. 3.2 Energy monitor. *Source* www.theowl.com

The Department for Energy and Climate Change [126] commissioned research to measure the public's views on smart metres in order to establish consumer awareness, understanding of and attitudes towards domestic smart metres. Half of energy bill-payers living in Great Britain had heard of smart metres (49%), with one in twenty claiming that they have one installed (5%). A third (32%) expressed support for the installation of smart metres in every home in the country, while one in five (20%) were opposed. Almost half (48%) of all respondents were undecided about smart metres. Support for smart metres were both highly correlated to age and size of household; with younger and larger households expressing greater support and interest.

The perceived benefits of having a smart metre installed included being able to better manage household finances (33%), to help avoid waste (26%) and produce a greater accuracy of billing (19%). Perceived disadvantages included cost (either to themselves, the taxpayer, the government or the energy companies) (19%) and data security (10%).

It was found that customers tend to passively receive smart metres rather than actively request them. Smart metre customers who refer to their display were generally positive about its impact on their electricity use and household finances. Almost six in ten (57%) bill-payers agreed that they felt more in control of their energy bills. In addition, almost seven in ten believed that the smart metre would help them reduce the amount of money they spend on electricity, and the amount of electricity they use in the household (69% each).

However, the DECC research showed not everyone who received a smart metre uses them; 44% of customers had either never looked at it (22%) or not installed it (22%). Only three in ten (31%) bill-payers expressed an information need around smart metres. This was higher among certain groups such as the middle-aged (40% for those aged 35–44), larger households (38% for households with 5 or more people) and those who do not speak English as their first language (38%). The most common information needs include functionality, costs and benefits. There is growing evidence that having a smart metre alone is not sufficiently habit forming for some people and therefore would not result in significant behavioural change as highlight by a recent Technology Strategy Board report [143].

Another study by Van Dam et al. [144] on the impact of home energy monitors explores the extent to which participants manage to sustain their initial electricity savings over time, with a particular focus on the development of energy-saving behaviour. The results showed that the initial savings in electricity consumption of 7.8% after 4 months could not be sustained in the medium to long-term. A second finding was that certain groups of people seem more receptive to energy-saving interventions than others.

The household that saved the most (a family of three) saved 42.6% of electricity during the pilot and 30.4% during the follow up. By contrast, the worst performer (a family of two) used 33.6% more electricity during the initial trial and 40.6% more during the follow-up. The reasons for these extremes were quite straightforward. The 'high spender' (initial electricity consumption of 3985 kWh/year) had just purchased an air conditioner, whereas the 'top saver' (initial electricity consumption of 2673 kWh/year) followed a strict regime: writing down the electricity metre data twice a day, recording how often the washing machine and dishwasher were used, replacing all incandescent bulbs, limiting the use of the tumble dryer to the bare necessity, decreasing the use of the dishwasher, and placing a timer on the pump of the garden pond. He made an extra note that he was 'very proud' of his achieved savings.

'Eco-spur' or 'Eco-steer' interventions and strategies encourage and facilitate the user to adopt resource efficient habits through the prescriptions and/or constraints of use embedded in the product or building design, making it easier for the user to adopt sustainable behaviour or reform their existing bad habits to a more sustainable lifestyle pattern. 'Eco-technology' and 'intelligent design' ensures behavioural change can take place by providing the tools, hardware or systems to enable that transformation to take place.

Suppliers of energy-efficient and green building products must be ready, willing or coerced into educating their consumers in the reasons for adoption. This is an added challenge beyond traditional marketing efforts [145]. Green consumers read labels, desire information and want control in their personal environments. For example, rather than remove the energy rating label from its appliances when they found that consumers equated lower energy with lower performance, a manufacturer chose to work with retailers and with consumers to address misconceptions about the efficacy of energy-efficient appliances and to educate people about their benefits, including the average 5-year payback period.

3.3 Psychology, Buildings and Sustainable Behaviour

Environment and Behaviour

Experiments by environmental psychologists, particularly expounded through Gestalt theory, a branch of psychology that emphasises personal responsibility and focuses upon the environmental and social contexts of a persons' life, the self-regulating adjustments people make as a result of their overall situation and their innate ecological wisdom, have shown that our environment is a determinant or influence on our behaviour [146, 147] and that behaviour and our environment mutually affect each other. For instance, the availability and price of energy sources determines consumption patterns whilst energy efficient behaviour determines the amount of pollution that will occur.

Steg and Gifford state that:

> Behaviour always occurs in the context of a physical environment (which) is crucially important to our thoughts, feelings, performance, behaviour and well-being and human activities also have an impact on the environment, through our behaviour, in terms of energy, water, vehicle and land use. To promote pro-environmental behaviour, we need to know which factors influence various types of environmental behaviour, and to develop and evaluate policies that successfully promote pro- environmental behaviour.
>
> (Steg and Gifford [90], p. 192).

Therefore, it may reasonably be argued that sustainable building design can influence people and their behaviour whilst equally, that the way people behave affects the sustainable performance of the building.

In terms of environment-behaviour theory, Zeisel [148] describes 'environment' as physical, administrative and social attributes of settings in which people live, work and play and 'behaviour' as things people do including thinking, feeling and seeing as well as talking with others and moving around.

Lewin's [147] model of behavioural change considers three phases. First, in order for change to occur and last over time, an explicit *unfreezing* process needs to take place, preparing people for change by challenging the barriers that inhibit change. The second consists of the *change* itself whereby the individual, unfrozen from past behaviours, is willing to consider alternatives. The resisting forces are likely to remain, and the individual is likely to continually reassess the desirability of change. Even after change takes place, it is still possible for the individual, organisation or institution to revert back to past unsustainable practices. Old practices still exist and can easily be adopted. New procedures are unknown, so they must be reinforced and *refrozen*. Under this model people, organisations and institutions are likely to be more susceptible to behavioural change and the adoption of sustainable building practices.

Mind the Gap!

The gap between our attitude towards urgent environmental and social issues and our actions or behaviours in tackling them can be applied to the design, construction, operation and use of buildings. Gaps have been identified, most notably by Leaman and Bordass [149], that exist between designed and as-built performance, known as the 'credibility gap' due to problems of communication between stakeholders, lack of integrated design practices between professionals, operational difficulties and unconsidered or unintended behaviours of building users.

Even the best efforts to develop an effective sustainable building will be undermined if, in its design, construction, operation and use, sustainable practices and technologies are not understood or made explicit at each phase. These schisms are known as the 'value-action' or 'intention-behaviour' gap by behavioural psychologists.

Paul Maiteny, an integrative and transpersonal psychotherapist, anthropologist and ecologist attributes the reasons for these gaps to a highly complex set of human responses to environmental, social and economic problems based on fear, social norms, social traps and other psycho-emotional blocks and barriers. Social, environmental and economic problems are symptoms of human problems that result from individual and collective behaviour and structures. These, in turn, emerge from particular cultural values, beliefs, interests and priorities [150].

Behaviour change is non-linear and non-rational and there is a disconnection between attitudes and behaviour. One study by Pichert and Katsikopolous [151] shows that though people stated a preference for green electricity their actual selections were based on the least effort by accepting the default offering. This recognises that people are 'boundedly rational,' that they behave rationally within certain limits but their behaviour becomes contradictory to their beliefs under certain circumstances.

Deeper psychological and emotional causes have been identified by Maiteny [150] that we need to engage with involving tapping into personal fears as a method in campaigning and public education just as emotional triggers are key to mainstream commerce's and politics' methods of convincing and seducing.

The need for society as a whole to change behaviour to a more sustainable way of being is widely accepted as we increasingly acknowledge that resources are finite, pollution levels must be controlled and the burning of fossil fuels impacts on the climate and global warming, with negative environmental, social and economic implications from increased risk of flooding, extreme weather events, biodiversity loss and ultimately the existence of life on Earth.

The Commons Dilemma

The 'commons dilemma,' first expounded by Hardin [152], explains how individual needs are pitted against those of the larger community, sacrificing individual advantages for environmental gain, and is primarily focussed on finite resource depletion and the exploitation of scarce resources: If we want the 'commons' to survive each of us needs to give up some of our freedoms.

The commons dilemma highlights the conflict between individual and collective interests. Applying this in a built environment context the dilemma may be perceived as being between the need for developers to exploit the limited natural capital, comprising of finite resources used in the production of buildings that make up our towns and cities, versus the need for society to develop in a more sustainable way.

People do not need to be aware of the conflict and it is common that people do not acknowledge or value the collective problems of their behaviours. In large scale commons dilemmas, individual contributions to collective problems and their solutions are often seen as futile. From currently available literature and case study research findings, property developers, architectural practices and construction firms that practice sustainable building methods tend to operate in a niche market, driven by significant social, environmental, moral and ethical concerns in response to the relatively limited demand for exemplary sustainable buildings and the majority of the mainstream built environment professions continue to consider themselves ineffectual in influencing global environmental problems.

Built environment professions comprising of individuals and organisations acting in an unregulated market, given the choice, tend not to sacrifice the economic benefits of non-sustainable behaviour. Pro-environmental behaviour is often associated with higher costs in terms of requiring more time, money or effort to achieve. For example, many architects view researching sustainable strategies as inconvenient when prescribed solutions are available. Contractors often believe that sustainable materials are more expensive and building managers believe sustainable and renewable operating systems are harder to manage or do not perform as well as traditional, fossil fuel-based systems reverting to organisational and institutional norms.

Lack of trust in others co-operation tempts individuals to act in their short-term self-interest, this is why many builders still build to minimum standards whilst allowing others to act more sustainably as they feel they would sacrifice their profit margins for no tangible gain. Thus, in the short term, the rewards for not engaging in pro-environmental behaviour are often greater than the rewards for engaging in it, no matter what others do. Individual choices are seen as having little impact and unsustainable practices are therefore perceived as excusable.

However, further evidence from literature review and the case studies, discussed in forthcoming chapters, show that some mainstream contractors are adopting sustainable practices because they see long-term advantages for their businesses as changes in consumer demand and corporate social responsibility develop.

So why then do individuals and organisations choose to engage in pro-environmental behaviour and not take the classical economic position of highest benefits (money, time, effort) against lowest costs? This mechanistic *homo economicus* view of human beings ignores the fact that people, organisations and institutions do tend to consider, at some level, what is the right thing to do, incorporating moral costs and benefits into decision making. This is partly explained by the 'norm-activation model '(NAM) from social psychology research.

Norm Activation Model (NAM)

According to NAM theory behaviour change occurs in response to moral obligations when participants are aware of adverse environmental consequences of their actions and when they believe they can reverse these consequences, e.g. specifying sustainable materials, technologies or using sustainable design strategies.

According to NAM it is more likely that sustainable choices are made when there is awareness of environmental consequences of unsustainable choices *and* it is thought that this will help reduce environmental problems. When both awareness and belief in positive consequences are high there is a strong moral obligation. If both are low then it is likely that unsustainable choices are made, resulting in unsustainable behaviour. NAM is more successful in explaining environmental behaviour when it is associated with relatively low behavioural costs in terms of money, time and effort. It has less explanatory power when the behaviour in question is more costly in terms of time, money and effort. The construction industry tends to fall in the latter category of this 'self-serving denial' because short-term high yield profit and gains are key, and likely to be the main driver for the development of the built environment as long as traditional social, economic and environmental norms prevail.

According to Steg and Gifford [90] there are four types of self-serving denial:

(i) People may disregard, distort or minimise environmental problems, e.g. climate change denial or disputing the efficacy of renewable technologies.
(ii) People can discount their liability for these problems believing their own contribution is small or ineffective and viewing the environmental problem as the result of collective rather than individual decisions and actions, e.g. it is the responsibility of authorities or the building industry to act sustainably.

(iii) Denial of ability or competence to perform pro-environmentally, e.g. lack of training or guidance in sustainable construction techniques.
(iv) Arguing that individual small scale sustainable actions would not be effective in reducing environmental problems. This defence mechanism is effective in the case of large-scale environmental problems such as the environmental impact of the built environment as a whole.

Habitual Behaviour

Habitual behaviour assumes that people and organisations think rationally before they act. However, in many situations often they do not. In the building industry behavioural norms and habits are historically strongly embedded and formed over many years of what is traditionally perceived as successful practice. According to Aarts et al. [153] habits have three important characteristics:

(i) They are activated in the presence of a specific goal, e.g. constructing a building.
(ii) The same course of action will be repeated when outcomes are satisfactory, e.g. the successful completion of a building. The more positive the outcomes of the action and the stronger the association between the goal and the action performed to reach that goal becomes, the greater the habit strength, e.g. building using the same design and construction strategies.
(iii) Habitual responses are mediated by cognitive processes. When people (and organisations) frequently act in the same way in a particular situation, the mental representation of that situation will be associated with the representation of the particular goal-directed behaviour. The more frequently a particular situation is associated with a particular behaviour, the greater the strength and accessibility of that association, and the more likely that an individual (or organisation) acts habitually, with the potential for repeated unsustainable behaviour.

Habits are a particularly difficult challenge for behaviour change initiatives, because they are largely automatic, but also triggered by the environment related closely to people and places. Habits or behaviours in themselves are not bad: the very same factors that lock unsustainable behaviours in (bad habits) will do the same for sustainable behaviour (good habits). A combination of legislation, enforcement, education and structural change plays a crucial role in this behaviour change process.

Verplanken and Wood [154] cite programmes in Canada that provide information to new residents about local products and services. Because this is during a period of ‘habit deconstruction’ the programme is more likely to succeed. Targeting points at which habits are likely to shift may be critical in maximising behaviour change initiatives. In terms of the built environment this could be during formal education for architects, engineers and tradesmen or during professional training, changing job or moving to a new office/factory building for occupants.

Steg and Gifford [90], p. 196 state that: ‘habitual behaviour is triggered by a cognitive structure that is learned, stored in and retrieved from memory when we perceive particular stimuli.’ In the case of the built environment professions the stimuli is often the formal education undergone and the organisational practices within which individuals operate.

Habits result in selective attention and a rapidly changing agenda can easily result in people not paying attention to information that might alter their choices. In general, habits are likely to be reconsidered only when the circumstances have changed significantly and within the construction industry education, training and changes in legislation offer some of the best opportunities for changes in personal and organisational habitual behaviour.

It may be concluded that when organisational and institutional habits are strong, reasoned choices become more difficult or are made impossible, therefore some of the theories of reasoned action discussed thus far are not wholly useful for explaining habitual behaviour. The construction industry and the way we design, build, operate and use buildings are habitually strong. Research by Ajzen [155] and Verplanken et al. [156] shows that when habits are strong, intentions are unrelated to behaviour.

However, existing habits may not always yield optimal outcomes particularly when the circumstances of the situation have changed. The sustainable construction agenda is rapidly changing, with a wide variety of drivers making sustainable and pro-environmental options more feasible and often preferable in terms of time, effort, money and corporate social responsibility.

3.4 Interventions and Strategies for Behaviour Change

Behavioural interventions and strategies can modify environmental behaviour and can be classified as either antecedent or consequent strategies.

Antecedent strategies target perceptions, cognitions, motivations and norms and include: education, verbal and written prompts, modelling and demonstrations and behavioural commitment procedures. This approach assumes that increased knowledge or positive attitudes will result in the higher adoption of sustainable and pro-environmental behaviour. Consequent strategies change the consequences that follow behaviour and include penalties, rewards and feedback. According to Steg and Gifford [90] behaviour analysts consider consequences to be the primary determinant of voluntary behaviour.

Antecedent Interventions

Education

Education and training is important in preceding behavioural change (pre-construction) providing a strong rationale for the desired change. This is why educational opportunities in sustainability through lifelong learning, that is formal and informal educational experiences to continuing professional development, are critical and relate strongly to sustainable behaviour change in the built environment disciplines. Training is different from education in that training usually involves a role-playing and feedback component.

The effectiveness of educational and training interventions can be enhanced by tailoring the information to the target audience. Information delivered interpersonally is more effective when it is done in small, rather than large, groups and when it actively involves participants in relevant activities and demonstrations. Educational interventions are discussed in Sect. 3.6 of this chapter: 'Built environment and pedagogy' and for the case study findings see Chaps. 6 and 7 and further discussion and conclusions in Chap. 7.

Signs and Prompts

Signs and prompts can be incorporated into a building to remind us we have attitudes which are favourable to pro-environmental behaviour. Prompts are cues that convey a message. Research by Becker and Seligman [157] has shown that their effect is relatively small but statistically significant and need to be experienced by many people over a long time frame.

Geller et al. [158] identified several conditions under which prompting strategies are most effective. Specifically, prompts work best when (1) the target behaviour is clearly defined by the prompt, (2) they are relatively easy to perform, (3) the prompt is located where the target behaviour can be performed and (4) the message is stated using appropriate language, e.g. politely.

Prompts can adopt intelligent design to target desired behaviours (see Fig. 3.3).

It is proposed in this thesis that an entire building can serve as a prompt throughout its lifecycle and particularly at post-construction stage if it is designed to target sustainable behavioural change of its occupants and visitors by making its operation and use as visible, tangible and interactive as possible. This requires a design team that has an understanding of behavioural change and pedagogical strategies or who collaboratively work with experts in these fields as part of an integrated design team as discussed in Chap. 4. They also need the drive and desire to produce buildings that are sustainable throughout their design, construction, continuing operation and use. And, perhaps more importantly, and conversely, that the majority of buildings that do not consider sustainability throughout their lifecycles have an adverse effect on

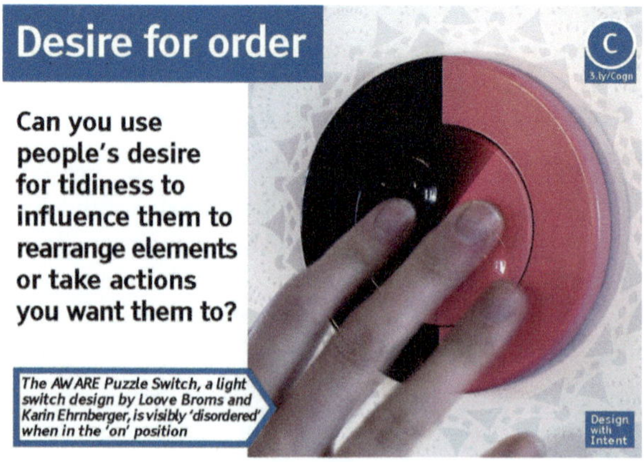

Fig. 3.3 The AWARE puzzle switch. *Source* Lockton [159]

sustainable behaviour reinforcing habitual unsustainable behaviours for designers, contractors, operators and users of buildings.

Modelling and Demonstrations

While prompts are appropriate for simple behaviours, modelling may be a more appropriate approach when the desired behaviour(s) are complex, such as the production of a building. Modelling involves demonstrating specific behaviours to a target audience. Research by Winnett et al. [160] suggests modelling may be more effective when the models are presented with rewarding consequences immediately after the desired behaviour is realised. Interactive scale models are a useful tool not only to help designers envisage how features inter-relate but also to demonstrate how a particular feature of a building works or how a whole building sustainable strategy operates, particularly when it invites interactivity and adopts experiential learning strategies. Figure 3.4 shows how a model is used to make clear the combined environmental features of a building. Physical models can also be used during occupation of the building to allow the users to visualise the systems that are installed. Sophisticated computer modelling can provide virtual representations of a building with explanatory, demonstrative and interactive qualities as well as providing feedback on the environmental and economic performance of the building.

Behavioural Commitment

Behavioural commitment is in effect a formal undertaking to perform a targeted behaviour, for example, to produce a sustainable building that enables occupants to

Fig. 3.4 Cut-away model of BRE environmental building in relation to the actual building (Fielden Clegg architects)

behave sustainably. Engaging with environmental assessment methods for buildings, such as BREEAM, are in essence a behavioural commitment to achieve a building that has a given level of performance according to a number of sustainable criteria. Commitment has been shown to have prolonged effects by Geller [161, 162] and the adoption of environmental assessment methods throughout the developed and developing world has been a strong driver for the sustainable construction agenda and has undoubtedly changed individual and organisational behaviours. As a commitment to build sustainably is generally in the public domain individuals and organisations tend to act in accordance with their assurance because they want to avoid the disapproval by others of their behaviour.

Behaviour analysts explain the tendency to follow through on commitments with the notion of rule-governed behaviour. Individuals and organisations learn rules for behaviour and through experience learn that following the rule is associated with positive consequences (reputational, financial, experiential) and breaking the rule is usually followed by negative consequences. Social psychologists such as Cialdini [163] attribute this tendency to follow through on a behavioural commitment to the powerful social norm of consistency, which creates pressure to be internally and externally consistent.

Consequent Strategies

Rewards and Punishments

Strategies will have limited effects when there are few opportunities to change behaviour within the industry or when sustainable building practices are less attractive than current industry norms that utilise unsustainable practices. In these cases, stronger measures are needed, or measures that change the relative attractiveness of behavioural alternatives that impact on a more societal level. The type of consequence strategies that impact on behaviour within the built environment professions fall into two categories, reward and punishments and can take the form of:

- **Pricing Policies** to reduce costs of sustainable buildings or increase costs of sustainable buildings.
- **Legal Measures**—laws that prohibit or encourage the use of certain materials, methods or technologies. Such measures are only effective if laws and regulations are accepted by the industry and enforced by the authorities, and violations met with sanctions.
- **The Availability and Quality of Products and Services** making environmentally harmful behavioural options less feasible or new or better behaviour options may be provided such as more sustainable materials and technologies on the market.

Penalties are believed to be less effective because they are more likely to result in negative emotions and attitudes. When enforcement or incentives are inconsistent, or cease, behaviours are likely to return to their previous state (e.g. Feed in Tariff scheme or fly tipping). Steg and Gifford [90] show that both rewards and penalties will be more successful when they are '*just*' sufficient to initiate behaviour change. For example, individuals engaged in the built environment professions are more likely to attribute their behaviour change to their personal or organisational convictions, whereas too strong rewards or penalties could result in people attributing their change in behaviour to the rewards or penalties themselves. Consequently, they may only have short term effects, i.e. as long as they are in place, because they will probably not have changed their attitudes, for example, where a contractor is obliged to adopt a sustainable method but does not understand, or has not been told, why.

Feedback

Feedback strategies involve providing information to participants about the frequency or consistency of their behaviours. Such data make the consequences of desirable behaviours more salient and increase the likelihood of behaviour change corresponding with the consequences. Feedback may be given about the extent to which one's behaviour changed (e.g. 'you have cut your energy bill by £5'), the consequences of this behaviour (e.g. energy or water savings), or on the environmental impact of behaviour changes (e.g. reduction of emissions of greenhouse gases.

Feedback strategies have proved successful in targeting home energy consumption which show modest but consistent energy savings [158]. In this way, individuals become aware of the relation between their behaviour and environmental consequences. This informs designers, users and operators clearly what they are doing and helps facilitate them to make environmentally friendly decisions through offering real-time feedbacks, for example, a thermo graphic image of a building element that is performing inefficiently can be directly visualised as heat loss and easily translated to cost and potential savings, encouraging or guiding the facilitator to act ecologically (see Fig. 3.5).

Feedback displays from renewable energy systems explicitly show reductions in carbon use and production of renewable energy which can act to incentivise individuals within organisations. The advent of intranet systems has allowed individuals to access the performance of their building down to individual workstations enabling them to access real-time data and follow the direct results of sustainable behaviour (see Fig. 3.6).

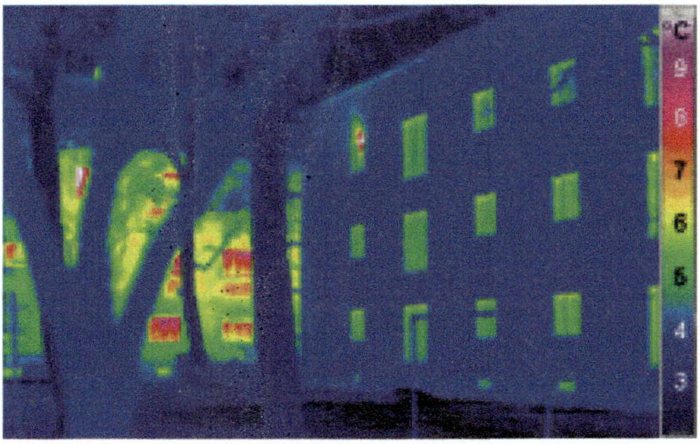

Fig. 3.5 Thermographic image. *Source* Archilab

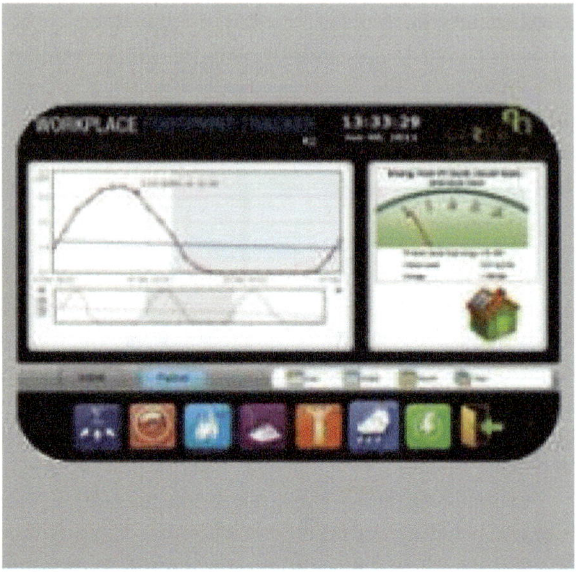

Fig. 3.6 Workplace footprint tracker. *Source* CEREB

It has long been understood, shown by research undertaken by Kohlenberg and Phillips [164] over 40 years ago that consequent strategies have proved better than antecedent ones (intervention *before* a given behaviour) in relation to pro-environmental behaviour through their research related to picking up litter through positive reinforcement techniques compared with prompts such as posters requesting litter-picking. Research by Van Houwelingen and Van Raaij [165] indicate that daily feedback of gas consumption is most effective in reducing household gas consumption (12.3%) compared to receiving information only (4.3%), being taught how to read the gas metre (5.1%) and receiving monthly feedback (7.7%).

It is clear that interventions and strategies can modify negative sustainable behaviour. It is proposed that sustainable buildings allied to information provision and education for sustainable development (ESD) can have a significant impact on environmentally responsible behaviour. The building itself can act as an antecedent strategy preceding the behaviour being targeted for change, e.g. attitudinal change, education and prompts. Geller et al. [158], Newholme [166] and Olsen [167] discovered from experimental research that general positive attitudes to the environment are not very predictive of eventual behaviour change e.g. 'I'm an environmentalist' whereas specific attitudes can be successful in predicting related behaviour e.g. 'I am conscientious about turning off lights.' Dwyer et al. [168], Heberlein [169] and Kempton et al. [170] found that simply educating people is not wholly effective and may result in weak attitudes that may revert back to conventional norms of behaviour.

Steg and Gifford [90] discovered that generally, interventions are more effective if they are combined. For example, feedback to architects on how their design performs through post occupancy evaluation is likely to be more effective when information on how to adopt sustainable practices is available at design stage, through prior training and education and the architectural practice commits to design and build sustainable buildings, developing a virtuous circle of pro-environmental behaviour. However, these interventions do not guarantee change, especially when habits are strong.

3.5 Psycho-Emotional Blocks and Catalysts

As social beings we are inherently subject to and highly influenced by social norms. As shown by Newholme [166] those who are well informed are more likely to adopt environmentally responsible views and views translate into corresponding behaviour.

What is now known as 'nudge' theory postulated by Thaler and Sunstein [171] raises serious questions about the rationality of many judgments and decisions that people make drawing on psychology and human fallibility as a key feature for understanding human nature and irrational unsustainable behaviour. This was predicted as far back as 1966 by the visionary philosopher, architect and inventor Buckminster Fuller:

> I made up my mind...that I would never try to reform man—that's much too difficult. What I would do was try to modify the environment in such a way as to get man moving in preferred directions.
>
> (in Kolbert [172], p. 11)

Another key aspect are 'social traps,' situations in which personal interests with short term focus e.g. driving a large car, conflict with societal needs e.g. tackling climate change, that have a long term focus. Personal gains appear to outweigh the social and environmental costs. Global warming and climate change may reasonably be seen as a risk. Risk is perceived as higher if it is seen as uncontrollable, inequitable, catastrophic and likely to affect future generations, risk is perceived as lower if activities are controllable, voluntary, individual and not globally catastrophic. Slovic [173] proposes that the problem is that the perception of risk is not the same as the risk calculated by experts.

One way to overcome the tendency for people to over-discount the future is to frame sustainable building around an appeal to protect the health and welfare of their children or grandchildren. Wade-Benzoni [174] argued that over-discounting occurred because the harms created were often far off in the future, uncertain, and affected people with little personal affinity. But relating this to one's offspring shortens that distance and makes the benefits more present than might otherwise be possible.

Individuals' impact on the environment is becoming increasingly powerful. With the exception of natural phenomena like earthquakes and volcanic eruptions the vast majority of sustainability problems are caused by the human species. Environmental problems like social and economic problems are symptoms of human problems as a result of individual and collective behaviour and structures. These, in turn, emerge from particular cultural values, beliefs, interests and priorities.

Deep psychological and emotional triggers are linked to ecological behaviour, identified by Maiteny [150] who illustrates how emotional triggers are used in mainstream commerce, advertising and political methods of convincing and seducing such as the need for insurance, social convention and the smoking ban. We need to engage with tapping into personal fears as a method in campaigning and public education just as emotional triggers are key to mainstream commerce's and politics' methods of convincing and seducing, e.g. Pensions sold as security for the future freeing you from worries today could equally be applied to sustainability issues.

Maiteny states that 'once a situation is recognised as a physical threat, such as global warming, then it is also experienced as a psychological and emotional threat' [150], p. 12. However, there is no guarantee that the response will be appropriate in terms of reducing the individual or organisational contribution to the problem through their subsequent behaviours.

Margai [175], p. 780 concludes that 'individuals who believe it will be difficult to carry out environmentally responsible behaviour are unlikely to engage in that action.'

A recent experiment on a popular television science programme [176], set up a hundred cyclists (Fig. 3.7) powering a mock-up of a 'typical' UK home inhabited by a 'typical' UK family. The family were invited to spend a weekend in the home and to behave as they normally would, which unbeknown to them was being powered entirely by the cyclists and monitored by a team of professionals. Professor T. Jackson from the Centre for Environmental Strategy, University of Surrey stated on the programme that the experiment was 'a good way to visualise what is essentially invisible to us.' He went on to state that:

> Energy is perceived as distant. We have emotional attachments to what energy does for us but little emotional attachment to or understanding of energy itself and related environmental and social consequences. By directly experiencing the consequences of the behaviours of others directly we can expect to change personal behaviour.
>
> (BBC [176] Interview with Professor Tim Jackson)

Fig. 3.7 The human power station. *Source* BBC [176]

On being shown the warehouse full of exhausted cyclists and some of the footage of their profligate behaviour their initial response was shock and surprise followed by statements that given the facts they would be more mindful of the need for appliances to be on at the same time, they would use appliances more efficiently and would purchase energy efficient appliances in the future. The presenter concluded that we don't think about our unsustainable behaviour until we see it on a human scale.

George Monbiot in a newspaper article stated:

> ....it is not that people aren't hearing about climate change, but that they don't want to know. The professional classes have the most freedom to lose and least to gain from an attempt to restrain it (climate change). Those who are most responsible for carbon pollution are—insulated by their money- the least likely to suffer its effects. It is also because it is consonant with the entire body of human self-deception. We want to be misled, we crave it; and we will bend our minds into whatever shape they need to take in order not to face our brutal truths.
>
> (Monbiot [177], p. 29)

Contradictory behaviour is known as 'moral (self) licencing' defined by Merritt as 'past good deeds that can liberate individuals to engage in behaviours that are immoral, unethical or unsustainable' [178], p. 345, such as using energy in a profligate manner because of prior investment in building insulation. The language we adopt often supports this contradictory behaviour with contradictory terminology such as 'sustainable consumption' which may legitimately be considered a contradiction in terms as current levels of consumption cannot ultimately be sustained.

It has been highlighted by Hoffman and Henn [11] that obstacles faced by the green building movement are no longer primarily technological and economic. Instead, they are social and psychological. It is often the behaviour of individuals, organisations and institutions that determine the level of engagement with the sustainable construction agenda.

Jenks and Dempsey [179] state '…it is behaviour, lifestyles and peoples' aspirations that are at the heart of achieving a sustainable environment. The form of urban areas and buildings within them, do not determine sustainable behaviour, but they might provide the right setting for it' [179], p. 8.

The development of buildings represents a complex set of processes and sustainability can often be seen as an extra layer of complexity by built environment professionals. From case study analysis presented in Chaps. 6 and 7 and evidence from other sources such as Frankiewicz it is clear that the built environment professions, and society as a whole, will have to make significant behavioural shifts in their organisational practices, partly driven by changes in mandatory legislative requirements and a moral obligation to act, if we are to achieve transformative advances in achieving a sustainable built environment.

The challenge is to focus on ways of developing environments that favour sustainable ways of life, opening up more opportunities for intervention and allowing climate change policy to benefit from a much wider range of social sciences, suggested by Professor Shove [180] as transition management, evolutionary economics and complexity science. These approaches look at complex blends of both internal barriers (attitudinal, opportunities, actual) and external barriers (contextual, abilities, perceived).

Of course, the whole premise of this study is based on the hypothesis that whole buildings can, and should, in their design, construction, operation and use transform human behaviour, through a myriad of interventions, to be mindful to act in more sustainable ways and the next section discusses how the built environment and educational opportunities can combine to achieve this.

3.6 Built Environment and Pedagogy

Introduction

Buildings have been designed, constructed and used throughout millennia and across civilisations to exalt the process of teaching and learning. The dreaming spires of Oxford and Cambridge are akin to some of the most religious architectural practices. In their grandeur, architectural edifice, consideration of spatial relationships and optimal utilisation of craftsmanship and materials they celebrate pedagogy through architecture aspiring to greater knowledge and understanding of the world, promoting and enhancing learning.

In the modern UK era with the advent of 'education for all' educational buildings have necessarily and pragmatically had lesser and humbler aspirations compared to the national monuments of their red brick counterparts. Through emerging Victorian values school, vocational and higher educational buildings displayed some of the elements of past achievements in terms of the use of durable materials, high thermal mass, good daylighting, controllable ventilation, a central heating system, generosity of space, amenity for play and the application of well-practiced construction skills (see Fig. 3.8).

The post-war years represented a move towards more austere and utilitarian designs and economic considerations meant the pedagogy of learning environments

Fig. 3.8 Victorian school building. *Source* www.geograph.org.uk

Fig. 3.9 CLASP school building. *Source* www.geograph.org.uk

was neglected in favour of mere space enclosure for the purpose of delivering education, culminating in some of the poorest performing educational buildings. These were exemplified by the Consortium of Local Authorities Special Programme (CLASP) system buildings of the 1950's, 1960's and 1970's. These were a method of designing and assembling prefabricated buildings for use in the public sector (see Fig. 3.9).

They were effectively a steel framed flat roofed box comprising of prefabricated elements made of steel and concrete with small windows placed high on the building, immediately below the roof. Wood or metal pillars held up canopies which were usually wood-panelled underneath. Pre-formed panels of aggregate-coated concrete formed the outer walls. The structures were cheaply built and quickly decayed and required almost constant maintenance and were therefore expensive to maintain in use. They incorporated asbestos, now known to be a carcinogenic trigger. By almost every sustainability indicator these buildings represented the antithesis of sustainable building, apart from the use of pre-fabrication, which is now extolled as a key sustainable strategy. They suffered from poor detailing, were prone to excessive solar gain in the summer and heat loss in the winter, the flat roofs were poorly detailed and subsequently leaked. Windows were single glazed with ill-fitting metal frames with poor ventilation control resulting in additional heat loss. There were concerns about fire safety, in particular the ease with which flames could spread through roof related spaces and the quality and effectiveness of acoustic insulation and the possible impact of noise disturbance on educational attainment. The Second Consortium of Local Authorities (SCOLA) system was similar to CLASP but used brick and timber instead of concrete.

System-building was successful in delivering the numbers of buildings required across the country and 10,500 schools had been built between 1945 and 1975. This was the height of school building in the twentieth century and a low point followed with the loss of power of the Labour government in 1979 which 'ushered in 25 years of neglect—the importance of the school environment ceased to be an issue' [181], p. 4.

Effective learning and teaching was reported to occur despite adverse physical conditions and lack of resources [182]. The First National Commission of Education Report Learning to Succeed made no reference to school buildings or architecture in its index. The follow-up study into the basis of improvement in ten disadvantaged areas indicated that while physical environment might not be a necessary pre-condition for improvement in some schools it was nonetheless an important and necessary condition for effective learning.

Pedagogical theories began to change and knowledge ceased to be just about what could be acquired and stored. This change in pedagogy influenced changes in school building design. Spaces were created to facilitate the ideal of the pupil as the agent of their own learning, development and social being. The link between school design and learning was apparent in the thinking of those advising the government and architects, that learning was both personal and individual.

The next significant movement in educational building design was the introduction of the very ambitious but contractually flawed Building Schools for the Future Programme (BSF) a £45 billion school building programme introduced by the Labour Party in 2004 with the ambitious target of making all schools zero carbon by 2016. Many BSF projects recognised the move towards collaborative teaching and learning strategies driven by educational change theory which reflected many of the aspects of theories of behavioural change, representing a synthesis of pedagogy and architecture. The pedagogical value of sustainable school buildings was recognised by the Prime Minister at the time, Tony Blair:

> Sustainable development will not just be a subject in the classroom: it will be in its bricks and mortar and the way the school uses and even generates its own power. Our students won't just be told about sustainable development, they will see and work within it: a living, learning, place in which to explore what a sustainable lifestyle means.
>
> (Blair [183]).

Despite these laudable aims the sustainability of many of the BSF schools did not meet these expectations largely through poor procurement practices, lack of change management and discrepancies between designed and in-use performance, dubbed the 'credibility gap.' Matt Bell from the government's architectural advisory body, the Commission for Architecture and the Built Environment (CABE) cites that: 'most of the design criteria for the schools was poor and mediocre' and that 'the buildings were no more energy efficient than a school designed 10 years ago' [184]. Architects were accused of being fixated with 'eco-bling' driven by standards that awarded points for green features that in use became over-complex 'nuisance' technology and were too focussed on aesthetics.

In their defence a lot of the technology was new for designers and local authorities and the projects viewed as testing grounds for learning, researching and developing more workable solutions. But with buildings learning by making mistakes is an expensive business. Many view the BSF programme with mixed views on its success but some of the responses of pupils indicate a pedagogical and behavioural effect:

'I'm happier to be in a more environmentally friendly school, I recycle more' and 'I never knew about the environment until I came to this school and it's made a big change in my life' [184].

Many of the building projects mirrored the integrated approach adopted by educationalists and in many cases utilised the ideas of teachers, students and behavioural psychologists in the design process. Changes in teaching strategies, driven by the increasing use of IT, blended learning and project-driven learning impacted on the spatial layout of buildings requiring non-linear configurations, away from the transmissive and didactic forms of teaching spaces to more inductive and transformative pedagogical strategies [185]. These arrangements fit well with some of the sustainable behaviour change theories previously discussed.

Figure 3.10 shows an exemplar BSF school which was conceived and created through a 'lengthy and rich' consultation process undertaken by the architects with one-on-one meetings with all department heads and workshops with teachers and pupils over a 12-month period. The Deputy Head is reported to have said 'what we really liked about the architects was they didn't have any fixed plans. The other architects all had pictures of other schools they had designed. The chosen architects said when you walk into this school it will be your school, not our school' [186], p. 5.

The Architectural Practice Director, when asked about what the success of the project was down to said, 'people who know about designing buildings got together with people who know about schools and talked to each other.' In the article about the school the reporter stated 'The consultation between architects, staff and pupils has led to a building loved by everyone' [186], p. 5.

Almost the first act of the new coalition government in 2010 was to axe the BSF programme altogether citing the PFI contracts as highly uneconomic and wasteful

Fig. 3.10 Minster school, Nottingham. *Source* Penoyre and Prasad LLP

and in the process abandoned much of the progressive work and lessons learned about sustainable building design and pedagogy. Arguably, the new administration have opposing educational ideologies in relation to the provision of public education and investment, soon bringing in the free school system which enables groups to set up schools in any existing building. The first free school to be set up was by the writer and journalist Toby Young, lauded as a pioneer by exponents of the free school philosophy, who stated in *Building Magazine Online*:

> Architects imagine educational buildings have a transformative effect on educational outcomes. But actually, that's bollocks! There is no evidence to support those claims at all. Educational outcomes are dictated by the quality of the staff, not the quality of the building. They are more likely to be inexpensive new builds. Crude, basic, meat- and-two veg. new builds that look like branches of Ikea. Bad news for architects!
>
> (Wright [187], p. 1)

Seven out of nine responses to this article strongly opposed Mr. Young's views, accusing him of 'trying to seek cheap and controversial public attention and publicity, to use unsuitable buildings without needing to meet educational standards and confusing free schools with cheap schools.' It was reiterated by a number of respondents that well-designed space does increase the learning ability of students. In the words of one respondent: 'In our brave new Britain a school should at least communicate to students that we as a society value education as much as we value shopping malls.'

Clearly the drive to provide buildings for pedagogy on a national scale is set back to some degree with the wholesale scrapping of the BSF programme, but that's not to say that sustainable educational buildings are not being constructed and some of the best examples are driven by enlightened local authorities and private developers working with sustainable developers, architects and contractors.

Once again in the UK, there is a move towards standardisation in the provision of new educational buildings, which is not in itself a bad thing as long as in their design, construction and operation careful consideration is given to the impact of the building on positive pedagogical, social and environmental outcomes. Baseline designs produced by the Education Funding Agency (EFA) resulting from the recommendations of the James Review indicate that new schools are set to be 30% cheaper to develop resulting in a £6 million saving on a typical BSF 1200 place secondary school and 5–15% smaller under the government's new standardised school approach.

The designs have simple orthogonal forms and are all based on the same limited range of room sizes and standard dimensions to enable a modular approach. Exposed thermal mass is used as part of the occupant comfort strategy and in terms of sustainability there is consideration for: a durable, airtight and well insulated fabric; maximum use of daylight while limiting solar gain; mechanical or hybrid ventilation with heat recovery in winter and natural ventilation in summer; and provision for the school and the contractor to monitor energy use. Energy efficiency savings have been estimated to be in the region of £70,000 per school each year.

Efficiencies are also expected to be driven through the use of building information modelling (BIM) technology and design for off-site manufacture and pre-assembly estimated to reduce on-site wastage by 50%. They are set to achieve BREEAM 'very good' as a minimum standard and each building will include a minimum 10% local renewable energy target.

In a recent article [188] RIBA president Angela Brady was critical of the government's approach, saying: 'Improvements must be made to the proposals to make sure that the schools we build now will suit the future generations of children that will learn in them, and deliver what the community needs in the longer term.' Clearly there is a danger of aiming to achieve minimum standards only, stifling the opportunity to achieve truly transformational school buildings.

There is a growing body of literature that rejects calls for the standardisation of buildings in the name of sustainability:

> There may well be as many types of relationship between nature and architecture as there are architects and buildings.
>
> (Jodidio [189], p.3).

Edwards [57] celebrates the fact that the agenda of sustainability is not leading to a single universal style but to a rich and complex architectural order around the world. Williamson et al. [20] assert that there is no class or style of design which is unequivocally sustainable architecture, and no fixed set of rules which will guarantee success if followed.

The evidence presented in this study strongly suggests that the design of school environments impacts on the health, well-being and productivity of pupils and staff. If the key determinant becomes cost at the expense of pedagogical, physiological, environmental and social benefits the school building movement will have taken a regressive step.

Buildings as Pedagogy

There is a plethora of sustainability-driven buildings being constructed around the world and many of the best examples have an educational and community function. Each have unique features and focus on widely varying degrees of sustainability, often embedding pedagogical and behavioural change interventions. Arguably, some of the best advances in interdisciplinary research, theory and practice are in the development of buildings linking sustainability, health and well-being and educational attainment. In their desire to promote sustainability, some go beyond the design and construction stages and attempt to embed features that have an educational and/or behavioural impact throughout the entire lifecycle of the building.

Paola Sassi in her book *Strategies for Sustainable Architecture* [190] highlights opportunities for buildings to promote awareness of sustainability and to use buildings as 'educational vehicles' are highlighted which also ensures more efficient operation of the building. Sassi states:

> An overall sustainable design strategy should include raising awareness about sustainability in general, educating the client and the project team about sustainable design approaches and influencing users about the operation of the building.

(Sassi [190], p. 88)

Sassi uses case studies to show how buildings can educate communities, how buildings can be used to demonstrate solutions to sustainable design problems and how buildings can also educate developers and buyers.

Architecture and education intersect throughout the lifecycle of a school building and the question here is how to make the most of this interaction with architects supporting education and educators contributing to the design process. Anne Taylor in her book *Linking Architecture and Education* speaks of 'interactive environments that serve as three-dimensional textbooks for learning and the incorporation of elements in the physical learning environment that will compel students to be responsible for their own intellectual growth and whet their appetite for knowledge' [191], p. 3.

This includes, users, operators and general visitors fostering an ethos of sustainability throughout the institution cutting across curricula and encouraging interdisciplinary practices which can have an impact beyond the organisation itself, 'stitching it' into the community. Sustainable school buildings are held up as best practice examples in educational, architectural, planning, construction and facilities management circles with the social, economic and environmental force to change attitudes and behaviours to effect transformational change.

Sustainable Education

Strategies for the transition to sustainable education have been well-researched and are not distinct or separate from the environment in which it can take place. Education for sustainability is an ongoing learning process which actively involves multiple stakeholders in change to achieve sustainability. It is sometimes also called Education for Sustainable Development (ESD), Environmental Education (EE), Sustainable Education (SE) or Learning for Sustainability (LfS).

Whole school development approaches reflect whole systems thinking and is one way of turning sustainable development into reality. The commitment to change values, attitudes and actions that reflect social and environmental justice, fostering strong links with local communities, developing pupil self-esteem and the professional development of teachers promoting more discursive and co-operative learning environments.

It has been established that the acquisition of knowledge about sustainability through education and training goes some way to informing our attitudes to our own relationship to environmental issues but research has shown that there is a very weak correlation between attitude and change in pro-environmental behaviour or lifestyle and there are many additional socio-economic factors for this, as well as solutions.

Sterling [185], a leading practitioner in ESD identifies three orders of change in order for change to take place. These are (i) *first order change*; making adjustments in the existing system and within accepted boundaries leaving basic values and beliefs unchanged and unchallenged, (ii) *second order change*; critically reflective learning, thinking about our thinking and (iii) third order change; redesign of the education system, a deeper awareness of alternative world views, a shift in consciousness.

Sterling goes on to highlight the differences between transmissive and transformative education (see Table 3.1).

According to educational change theory [192] transformational change needs to occur across technological, cultural, micro-political, biographical, structural and socio-historical aspects of education. Sterling supports this and states that 'schools and organisations need to engage in deep change in order to facilitate deep change and they need to transform in order to be transformative' [185], p. 15.

Sterling [185], pp. 60–61 describes a range of educational responses to sustainability:

1. **Education About Sustainability**—This has a content/knowledge bias and can be quite easily assimilated within the existing educational paradigm. This is an adaptive response, which equates with first order learning.

Table 3.1 Differences between transmissive and transformative education

Transmissive	Transformative
Education for change (practice)	
Instructive	*Constructive*
Instrumental	Instrumental/intrinsic
Training	Education
Teaching	Learning (iterative)
Communication (of 'message')	Construction of meaning
Interested in behavioural change	Interested in mutual transformation
Information—'one size fits all'	Local and/or appropriate knowledge important
Control kept at centre	Local ownership
First order change	First and second order change
Product oriented	Process oriented
'Problem-solving'—time bound	'Problem-reframing' and iterative change over time
Rigid	Responsive and dynamic
Factual knowledge and skills	Conceptual understanding and capacity building
Education in change (Policy)	
Imposed	*Participative*
Top-down	Bottom-up (often)
Directed hierarchy	Democratic networks
Expert-led	Everyone may be an expert
Pre-determined outcomes	Open-ended enquiry
Externally inspected and evaluated	Internally evaluated through iterative process, *plus* external support
Time-bound goals	On-going process
Language of deficit and managerialism	Language of appreciation and cooperation

Source Sterling [185], p. 38

2. **Education for Sustainability**—This includes content but goes further to include values and capability bias. This involves some reformation of the existing paradigm to reflect more thoroughly the ideas of sustainability, but otherwise the existing paradigm remains largely intact, for example with respect to unqualified economic growth. The greening of schools and colleges movement is largely located here. This is an adaptive response that equates with second order learning.
3. **Education as Sustainability**—This is a transformative, epistemic, learning response by the educational paradigm, which is then increasingly able to facilitate a transformative learning experience. This position emphasises process and the quality of learning, which is seen as an essentially creative, reflexive and participative process. The shift here is towards 'learning *as* change' which engages the whole person and the whole learning institution. This response is the most difficult to achieve, particularly at institutional level, as it is most in conflict with existing structures, values and methodologies, and cannot be imposed. This is third order learning and change.

Sustainable education concepts can be woven into the structure of a school, making it an active rather than a passive space for learning. The 'informed' learning environment is a three-dimensional textbook or teaching tool intentionally filled with rich prompts or cues for learning. The environment in which ESD is delivered can have an impact on the learning process, for example through experiential learning about sustainable construction practices, materials and technologies, in situ, particularly related to vernacular forms of architecture and the use of regional craft skills and local materials which is a fundamental aspect of the hypothesis proposed in this study, that sustainable buildings can, and do, influence sustainable behaviour.

Sustainable buildings can enhance ESD in the built environment sector leading to behavioural change in graduates and established professionals, as well as those using buildings who are not necessarily directly involved with the structure or operation of the building, but who benefit from occupying a sustainable building. Janda [12] recognises that although few students will ever become practising design professionals, all students use buildings and will continue to do so throughout their lives and it is argued that in order to prepare people to accept more responsibility for their role in the built environment, education should be much more comprehensive, hands-on and iterative.

Findings from this study, presented in Chaps. 6 and 7 and discussed in Chap. 8, show that education and awareness-raising via the built environment includes strategies to inform building designers, developers, contractors, managers and users of the consequences of their actions and encourages and guides them to change, such as on-site training or well-written operating manuals. This imparts 'eco-information' (as discussed at the beginning of this chapter) to be imparted offering behavioural choices, enabling the target audience to visualise or experience how their action impacts on the consumption of resources such as materials, energy, water, waste etc.

For educational and community buildings to be truly transformative, educators, building managers developers, engineers and designers must enter into meaningful dialogue at the conceptual stage of the building. An example where this has not happened was the introduction of a biomass system at a school. Teachers recognised it as a valuable teaching and learning resource but the system was so designed that it was very difficult to experience the system in operation because of siting, spatial limitations and poor access. If this had been considered at design stage its educational value could have been optimised.

Professor David Orr in his book *Nature of Design* [1] speaks of the 'hidden curriculum' that is the building itself and its inherent value in its potential to educate and raise awareness and bring about pro-environmental behavioural change. School buildings are also often referred to as the 'third teacher' and Ty Goddard, Director of the British Council for School Environments (in [193]) suggests that we ignore the built environment (as a teaching and learning resource) at our peril. Orr states 'other than as a collection of buildings where learning is supposed to occur, place has no particular standing in contemporary education' [1], p. 128. He argues that our sense of place has become nebulous and homogenised, our immediate places are no longer sources of food, water, livelihood, energy materials, friends and recreation.

Bonnett speaks of the 'relative invisibility of nature' [194], p. 711 in predominantly modern western lifestyles and poses the questions what is nature and what is our place in it? How can we know nature? What should our attitude towards it be? and against what criteria should humankind judge its progress/success/well-being in relation to the natural world? Such questions are pedagogically relevant because they represent important ways of articulating our understanding of the human situation. Such questions are central to addressing the environmental predicament we face today; the meaning of environment and nature, its importance to education and what taking sustainable development seriously suggests for the aims and curriculum of schools and their built form.

Firth and Winter [195] argue that schools are in danger of becoming epistemologically and socially irrelevant unless they recognise the nature of knowledge itself and the dynamic and creative processes required to translate this into teaching and learning situations. Thus, there is a need to reconceptualise the relationship between knowledge and pedagogy and that the aim of education should be to develop students understanding of the produced nature of knowledge. Essentially, to move away from teaching young people about the pre-existing world, a way of thinking/being/doing that is decided in advance leading towards a closure but to consider alternatives that complicate the scene, unsettling actions and understandings and demanding the exercise of critical choice. As the world becomes more complex and uncertain these skills are far more valuable in creating a more sustainable future. Edward Mazria states that:

> As the world transforms in its efforts to avert dangerous climate change, ecological literacy will become a vital part of all education. We cannot expect future generations to value what they are being taught, unless they also experience and live it in their learning environments. Surrounded by sustainable buildings, students will come to think of sustainability as being as obvious and essential to a building as doors and windows.'
>
> (Mazria in Taylor [191], p. 1)

Janda asserts that,

> few people other than architects are ever taught to read the language of the built environment. As a result, the general population tends to treat buildings as static objects rather than dynamic systems. By re-engaging with our immediate surroundings, we inherently become more aware of ecological, social, economic and political imperatives.
>
> (Janda [12], p. 19)

Goddard states that 'architects and educators need to become bilingual in order to communicate the needs of modern schools across different disciplines.' He goes on to say, 'what you currently get is builders talking about building, architects talking about architecture and educators talking education' (Goddard in Firth [193], p. 1). Here we find clear recognition of the need for interdisciplinary working practices.

The function of a school building has significant impacts on the level of behavioural change that can be affected, although it may be asserted that all buildings should utilise their sustainable credentials for educational purposes. This study looks particularly at community and educational buildings and their potential for education, raising awareness, and experiential learning (embedded learning) for behavioural change. Education *about* sustainable buildings, landscapes and lifestyles can merely be an awareness lesson or theoretical discussion whereas education *for* sustainability uses (un)sustainable buildings as an educational resource to achieve a greater understanding of social, environmental and economic sustainability.

Linking the design concepts, fabric and operation of the building to what is taught within the building is paramount. Advanced pedagogical research from the National Training Laboratories Institute for Applied Behavioural Science [196] has long shown that experiential learning, or 'learning by doing' has much more potential for deeper, embedded learning than simple didactic methods, 'learning by telling.' Fig. 3.11 illustrates average student retention rates for a variety of teaching and learning strategies.

The learning pyramid (Fig. 3.11) illustrates that a student may learn about generic sustainable buildings through a lecture, by reading about it or through audiovisual stimuli but can only hope to retain up to 20% of that information whereas, using a

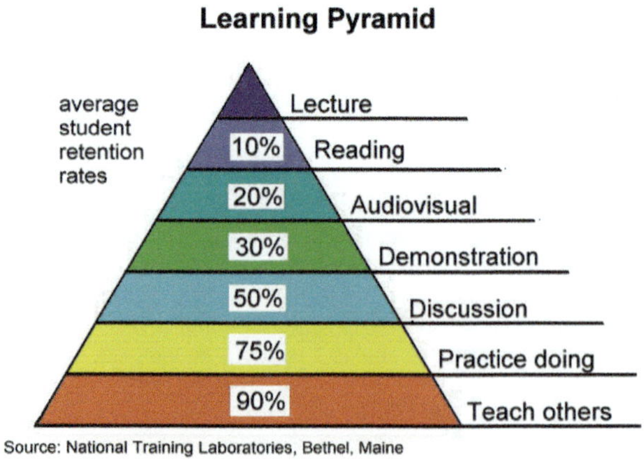

Fig. 3.11 The learning pyramid

building for demonstration purposes with discussion will retain between 30 and 50% and if you involve students and staff in the actual design, construction and use of the building you can expect learning retention rates of up to 75% enabling greater processing of that learning into actions and behaviours.

Buildings provide opportunities for transformative teaching and learning experiences and opportunities for experiential engagement and activities across the curriculum at school level and across many disciplines at graduate and professional levels. Built environment education is similar to arts and museum and gallery-based education in that it can challenge pupils with open-ended, creative learning tasks which can help increase confidence, enhance pupil motivation, provide work satisfaction, and contribute to the development of lifelong learning skills. We are surrounded by buildings and designed spaces and built environment education can provide learning opportunities and out-of- classroom learning experiences in the local community that are accessible, inexpensive and familiar to both teachers and pupils.

Dwyer [168] illustrates that simply educating and informing people is not wholly effective and environment is critical in facilitating behaviour. Newholme [166] also concludes that general positive attitudes to the environment are not very predictive of eventual behaviour. Factual knowledge can lead to a change in attitude towards behaviours linked to social and moral values that lead to subjective norms (socially appropriate action) and will eventually lead to behavioural intention and outcomes; this is illustrated by the work of Fishbein and Ajzen [197].

Architect Robert Kobet [198] suggests that secondary school facilities should be designed to function as an extension of the curriculum. School grounds that include gardens could provide participatory learning opportunities (as well as physical inputs) to school cafeterias, culinary classes and biology courses, especially as children are experiencing ill-health and behavioural problems as a result of indoor and sedentary lifestyles, suffering from 'nature-deficit disorder' a hypothesis proposed by Richard Louv in his book *Last Child in the Woods.*

No single discipline fully explains our world. Through interaction with objects, children encounter integrated concepts from mathematics, science, social studies, physical education, and the arts. As children discover the richness of the environment, they learn to view their surroundings in multiple ways, increasing their problem-solving abilities with ideas and strategies taken from all subject matter areas. Davis et al. [199] show that interdisciplinary learning automatically aids higher order thinking skills by requiring learners to compare, synthesise, and innovate.

Pallack et al. [200] show that attitudes formed from direct behavioural experience tend to be stronger and are more predictive of later behavioural change than are passive or abstract attitudes. Therefore, the more we use buildings as an experiential teaching and learning resource the more likely pro-environmental behaviour is to occur leading to resource efficiency, waste minimisation, increased health and well-being, enhanced biodiversity, low-impact transportation etc.

At a recent conference *A Roadmap to Sustainable Schools* held at the University for Central London by building engineers, a head teacher stated that architects and engineers lacked an understanding of the curriculum when designing and constructing educational buildings and during another presentation at the same conference an engineer stated that teachers and other educational staff do not understand buildings and in particular the function and purpose of sustainable buildings.

Herein, partly lays the answer. If both are engaged in the process of design and use of a building, bringing their respective knowledge of architecture and pedagogy together, the educational value of a sustainable building can be optimised. One of the legacies of a well-designed sustainable building will be its educational value and its incorporation into the curriculum in terms of its structure, the materials used its use of resources, its role in the community and ultimately its effect on behaviour of students, staff, parents and the wider community.

It is argued that sustainable buildings allied to sustainable education can have a significant impact on environmentally responsible behaviour through combined technical and pedagogical interventions embedded into the design, construction and operational processes. Full scale buildings designed for sustainability can also act to demonstrate sustainable materials, technologies and methods.

Therefore, education and learning as sustainability must be linked with effective learning spaces and is critical in fostering ESD. The theory and practice of creating effective learning spaces can be reflected in the building's spatial configurations as well as the materials, methods and technologies from which the building is constructed.

A best practice example of the potential for future sustainable school design is the 'Re-Thinking School' project (see Fig. 3.12) showcased at the Building Research Establishment Innovation Park in Garston, Watford. It combines pedagogical objectives such as education transformation and better learning outcomes with practical delivery of environmental sustainability to encourage 'sustainable evolution.' The building acts as a research tool for knowledge re-investment and is being used to generate case studies of both hard performance data and user experience data.

Key features and benefits of the Re-Thinking School building include:

Fig. 3.12 The 're-thinking school' building. *Source* BRE architect: white design

Fig. 3.13 Modular timber panels. *Source* BRE

- Solid timber panel construction made from off-cuts of waste wood (see Fig. 3.13).
- The central teaching hub can be configured to suit different learning activities: electronic, individual and group learning.
- A 'daylight chimney' brings natural light and ventilation from the roof area two levels above.
- The IT server room is naturally cooled using cool air from under the building.
- A combined biomass and air-source heat pump provide renewable heat and hot water.
- An educational window panel contains recycled plastic bottles providing additional insulation (see Fig. 3.14).
- The WC is directly linked onto the learning area to reduce the risk of bullying and other anti-social behaviour.
- A dynamic lighting system mirrors the spectrum of sunlight throughout the day to boost and aid levels of concentrations in pupils.

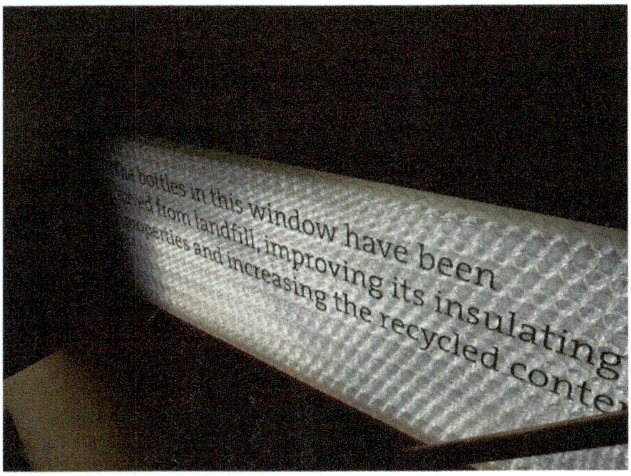

Fig. 3.14 Plastic bottle window. *Source* Author

- The green roof space maximises the usable footprint of the building providing external learning space, additional insulation, water attenuation and a habitat for biodiversity.
- Circulation areas, such as corridors, are fitted with seating and library area.
- A wind turbine provides zero carbon electricity and is an educational resource.
- Internal CO_2 and temperature levels and automated adjustments of openings are monitored by the weather station and controller sensors.
- The school is predominately clad in British White Cedar and the back of the building shows a number of sustainable alternatives, including used recycled mobile phones (see Fig. 3.15).
- Displays of construction system feature cut-a-ways and 'vision panels' show elements of the buildings structure and operating systems (see Fig. 3.16).
- Underfloor heating reduces energy consumption. The low emitter temperature levels lend itself to use in conjunction with a heat pump.
- Movable walls can break up or enclose space or be used as an object for projection and display.

The design recognises that students should be encouraged to progress at their own pace, consistent with their own skills and talents.

Fig. 3.15 Rear view of building. *Source* Author

Fig. 3.16 Transparent downpipe. *Source* BRE

The teaching spaces offer the ability to instruct a few pupils or individuals on similar topics at different paces using different learning strategies in order to allow for the customisation of each students personal profile.

Space and Layout

The spaces provided within a school and the relationship they have with each other contribute to their effectiveness in providing education and the efficiency of movement within the building. A study of 24 schools in Georgia, USA [201], concluded that the achievement of 8–9 year olds was most affected by circulation patterns and movement and 9–11 year old students by large group meeting areas. The layout can also affect aspects of sustainability and, in particular, the availability of daylight, options for natural ventilation and acoustic performance. A study of two new classrooms designed by INTEGER highlighted positive impacts on pupils' behaviour from the new learning environments that enhance mobility, flexibility and use of technology [202].

As educational needs and theories change so must design. Schools are moving away from didactic teaching styles that rely on uniformity, memorisation and lecture learning and into the digital information age and more towards interactive learning.

The old 'assembly line' model no longer supports what we know about how the brain/mind learns [203]. This means architects must provide different configurations for learning environments, more flexibility, adaptability, movable components, and future conversion to other uses. Research by Locker and Olsen [204] shows that poorly designed spaces can provide opportunities for social friction, bullying and other anti-social behaviour.

There are indications that the amount of space pupils have has an impact on both performance and behaviour, particularly among pupils with special educational needs. A study conducted by O'Rourke Swift [201] on elementary school pupils concluded that pupils performed significantly better in a number of tests and had improved behavioural patterns where the amount of space (indoor and outdoor) was above 9.29 m^2 per pupil. The new standardised school approach is reducing the amount of space available per pupil.

The benefits of small group working have been documented in a large British study of 10,000 students from 500 classes in 300 schools from Nursery age to the end of Key Stage1. The study concluded that when pupils were divided into groups in the classroom those in larger groups suffered greater adverse effects on their work and concentration [205].

Natural Materials

The use of natural materials has a number of environmental, social and economic benefits in terms of sustainability. It offsets the use of man-made materials which often contain volatile organic compounds (VOC's) and reduce or impair indoor air quality (IAQ) contributing to sick building syndrome (SBS) or building-related illness. Natural materials therefore have a positive impact on the health and well-being of occupants often improving productivity in work and learning environments. They are often locally sourced and in many cases are a biproducts from other industries reducing the environmental impact through transportation and processing resulting in low embodied energy and can stimulate local economies. Using the narrative of the lifecycle of materials can be a valuable educational strategy. At end-of-life natural materials tend to be more biodegradable and readily reused or recycled.

Embedding Sustainable Features

Through early design considerations sustainable features can be embedded that not only clearly show how systems work, enabling the building to 'explain itself' but can have additional educational benefits. For example, water management systems, consisting of green roof—rainwater distribution—wetlands filtration, can be made visible and offers a tangible narrative of how buildings impact on water use with issues such as scarcity, flooding, excessive run-off, extreme weather conditions through climate change and how buildings can be designed to mitigate these effects. By combining form and function in a sustainable way, in the words of the architect, the 'ethic is rendered aesthetic.' By taking an object such as a building, examining its form and function, and then translating perception of the building into ideas about sustainability, knowledge, understanding, creative problem-solving, and self-expression is encouraged.

Transparency

By making the operation of the building transparent and accessible, in terms of its structure, space, services and operation, occupiers and visitors can have a greater appreciation of how the building functions. Buildings can be designed to enable people to observe and participate in maintaining and caring for the building, through experiential learning, reflecting the need for 'architectural honesty' if we are to engage and educate people in the buildings they occupy. Common architectural elements can be designed for their maximum potential as learning resources to stimulate enquiry and learning.

Often people feel remote from the over complex and out-of-scale buildings that are imposed upon them whereas sustainable buildings engage, empower and instil a sense of being part of a whole. The concept of 'baubiologie' or building biology shows the need for us to appreciate buildings as living systems and the analogy can be extended to circulatory systems, boiler as heart, building management systems as brain external structure as third skin, breathability, sick building syndrome etc.

Impacts of School Environments on Learning

An increasingly broad body of research, presented below, shows a strong correlation between key educational outcome indicators and key sustainable building design characteristics.

Evidence suggests that school design affects the performance of a school, including behaviour, well-being and achievement of pupils, as well as the recruitment and retention of teachers, and the sustainability of the school itself. A survey conducted by the Teacher Support Network and the British Council for School Environments showed that teachers overwhelmingly agreed (95.8%) that the school environment had an influence on pupil behaviour [206].

The National Foundation for Educational Research carried out a study at Bristol Brunel Academy, a BSF school which opened in 2007, reported by the Commission for Architecture and the Built Environment (CABE). This showed that the number of pupils who said bullying was an issue for them had dropped by 23% compared with the school it replaced. Vandalism had dropped by 51% and the number of pupils who say they 'feel safe' had risen by 30% [206].

Various US studies of sustainable school buildings have revealed that student attendance rose by 5% after incorporating cost–effective indoor air quality improvements [207] better school facilities can add 3–4% points to a school's standardised test scores [208] and a 15% reduction in absenteeism [209]. Students moving from a conventional school to a new sustainable school achieving a LEED Gold rating experienced substantial improvements in health and test scores. A PhD thesis on the school found a 19% increase in average Student Oral Reading Fluency Scores when compared to the prior, conventional school [210].

In the UK Ofsted [211] visited 14 schools over a 3-year period and found that their focus on sustainability had a wide range of positive consequences. Sustainability captured the interest of young people because they could see its relevance to their own lives and futures. There was evidence of an increase in knowledge and understanding of the importance of leading more sustainable lives, and there were examples of more positive attitudes to learning, better behaviour, attendance, and improved standards and achievement. Importantly, the findings show that sustainability is a significant factor in improving teaching and learning in these schools.

In a study of 56 schools for the National College, Birney and Reed [212] found that schools which focus on sustainability bring an ethic of care and a common vision for building a more just and inclusive school and society. This results in positive benefits for young people's learning and well-being, and also for the staff and wider community. The study also found that a commitment to staff development through training and support enable teachers to understand a sustainable school's approach and how to make a school operate sustainably.

Research by the Sustainable Development Commission and by Thomas and Thomson [213] shows that environmental quality and young people's well-being are inextricably linked, and that young people's everyday experience of living and learning in the environment, and the health of the environment itself, are critical to overall well-being. The findings also show that the worse a local environment looks, the less children are able to play freely, and develop the habits and commitments that will enable them to address environmental problems in the future.

Hicks and Holden's [214] research into young people's views of the future, found strong evidence that, regardless of age, the environment is a consistent theme in their concerns about the future and that providing collaborative, positive and supportive learning environments is vital in helping students to raise and deal with their concerns. The research also found that young people become increasingly worried and/or disinterested when schools place too much emphasis on problems. Research from the *Cambridge Primary Review* noted that 'pessimism turned to hope when (people) felt that they had the power to act…the children who were most confident that climate change need not overwhelm them were those whose schools had decided to replace unfocussed fear by factual information and practical strategies for energy reduction and sustainability' [215], p. 198.

Gayford [216] found that where students were involved in monitoring, recording and reporting the effectiveness of the measures taken to improve sustainability within the school, or in planning changes in the school or local community, there were valuable educational outcomes, social networking, and increased student motivation. Gayford also found clear evidence from young people that telling them what to think and do about environmental issues is not effective. This and other studies demonstrate that using active, participatory and collaborative approaches helps young people to enjoy and achieve and enables the transfer of learning to everyday life.

A Building Research Establishment paper: *Health and Productivity Benefits of Sustainable Schools A review* [217] asserts that the positive health and pedagogical effects demonstrated by improved indoor air quality, lighting, acoustics, thermal comfort and spatial considerations are highly suggestive of a causal relationship between sustainable schools, high quality learning environments and improved educational and workplace outcomes. The review concludes that only by bringing a wide variety of datasets together on a significant scale can we draw quantified conclusions about the educational case for sustainable schools.

A recent pilot study by the University of Salford and architect Nightingale Associates reported in Building4Change [218] investigated seven primary schools with differing learning environments and age groups. The study involved collecting academic performance data from 751 pupils in maths, reading and writing at the start and end of an academic year and evaluating the holistic classroom environment, taking into account classroom orientation, natural light, colour, noise, temperature and air quality. Classroom design and use factors were included to identify what constitutes an effective learning environment. It was found that the classroom environment can affect a child's academic progress by as much as 25% over a year.

From evidence presented in this section it is reasonable to conclude that a school specifically designed to be healthy, and characterised by sustainable features such as optimal daylighting, fewer toxic materials, improved natural ventilation and acoustics and improved light and air quality would provide a more productive study and learning environment. It provides an opportunity to explicitly teach and learn about sustainability and therefore to potentially bring about sustainable behavioural change through experiential learning, bringing together buildings, pedagogy and people in a sustainable environment, that would not be present in a poorly designed school.

As presented, the building is a key factor, but other strong determinants include the motivations and priorities of individual teachers and their willingness to engage with the sustainability agenda in their subject area and the wider curriculum, set by organisational and institutional bodies at both local and central levels.

Critique of Education in Built Environment Disciplines

Research by Ryghaug [219], a political scientist, analysed the role of sustainability in architectural education using interview techniques, studying architectural competitions and trade journals. The major findings of the study were that sustainability on the whole is not successfully translated into something that is sufficiently important to most architects for them to give priority to these issues when difficult and conflicting decisions are needed.

The research also asserts that the aesthetic interests of architects are not effectively considered in the alternative eco-architecture and sustainable agenda. Sustainability gets 'lost in translation' when various interested groups try to build their networks because architects tend to create their own space between the networks, that is the one constituted by aesthetics. The values and actions of architects require complex and difficult choices in design and construction processes that are guided by more or less successful translations. Fallan [220] suggests sustainable approaches have a preference for instrumental and functional aspects rather than aesthetics and arguably do not fit well with contemporary views of form over function.

George Monbiot, author, academic and environmentalist, in a newspaper article *'Our craving for deception'* controversially stated that 'the only people I have met over the past few years who haven't the faintest idea what man-made climate change is, are university graduates' [177], p. 29. Although a rather sweeping generalisation, even if partly true it is a damning indictment of our current educational system that we still produce graduates increasingly focussed on narrow subject-based degrees that do not include education about one of the biggest threats to our existence, climate change and global warming.

Often sustainability is seen as a branch of the built environment delivered in a modular format that can suggest to students that it is an optional element rather than covering all aspects of design, construction, operation and use. This dis-integration makes it difficult for undergraduates and graduates becoming professionals, whose decisions directly affect environmental, social and economic sustainability, to break away from embedded theoretical norms that can preclude or sideline sustainability as an add-on. Much criticism aimed at sustainable architectural solutions is that they are merely superficial gestures with minimal impact, commonly called 'eco-chic,' 'eco-bling' or 'greenwash.' Apart from being technically ineffective these tokenistic interventions can convey a negative learning experience, e.g. poorly oriented photovoltaic panels, installed mostly for aesthetic reasons will produce low levels of electricity resulting in the perception that solar power is ineffectual.

Built environment professionals are highly influenced by the technical, social, organisational, financial, legislative and behavioural norms and rules introduced in their early training experiences. One approach is to integrate environmental education into existing curricula in the built environment sectors. This would involve architecture and engineering curricula in the university, apprenticeships in the building trades, and even training of owners and building managers [221]. There are a growing number of sustainable construction training courses emerging in programmes related to architecture, engineering management, urban planning and environmental issues.

Egri and Pinfield [222] conclude that regrettably, many do not foster cross-disciplinary collaboration necessary for a holistic approach to understand the relationship between the built environment and the natural environment. However, professional bodies are increasingly becoming engaged in driving the agenda through the accreditation of courses requiring sustainability in the curriculum. There is also a growing market in sustainability-related continuing professional development (CPD) courses for post qualification professionals, increasingly offered by mainstream organisations such as the Building Research Establishment (BRE).

A limited number of universities are incorporating sustainability on a more holistic level and this is true of some of the courses related to sustainable architecture aimed more towards, and attracting students from a wide variety of disciplines, as witnessed by the postgraduate courses offered by the Centre for Alternative Technology allied with University of East London. Students work in multidisciplinary teams and report surprising success in learning the basic assumptions and cultures of other disciplines allied with sustainability. Education can also take place with current professionals practising in a variety of fields.

The traditional role of the architect in the process of creating buildings has changed significantly from translating the promoter's brief into a finished building to effectively becoming a design sub-contractor with a loss of influence over many of the project outcomes. Naturally, architectural education reflects and potentially drives these changes as industry demands change.

Walker [223] criticises architectural studio design teaching as it centres on the individual and is characterised by the absence of a real-life collaborative design approach and that teaching focuses on the high-profile grand scheme rather than the modest commercial buildings with realistic budgets and technologies, which are the staple of most practising architects. The pedagogic justification for this is that if you can do the difficult design work you can, by default, do the easier work.

Walker [223] also asserts that a key characteristic of school studio projects are the lack of consideration for concerns a client might have in relation to cost (viewed as a construction matter), value (viewed as a property matter), return on investment or commercial risk. He cites a lack of teaching in the economics of the built environment which results in architects in practice who are not capable of engaging in meaningful way with the property and construction industries.

He also cites a lack of collaboration, leadership and team-building training when working with key project partners, particularly those involved in cost and value matters and argues that the architect must be trained to appreciate when critical life-cycle cost information should be used. Discussions about sustainability without a supporting commercial case are of little use in determining, or selling to clients, a particular design strategy.

Generally, an architect's training does not equip them to be able to demonstrate, both by design and financially, that spending more money on a building can be a good commercial decision as well as having better environmental and/or social dimensions. Although the Royal Institute of British Architects (RIBA) Part 3 professional studies syllabus does require an 'understanding of the social and economic context for investment in the built environment,' the bulk of the syllabus remains concerned with the management of architects, the management of the building process and construction law.

The relative global success of British architects has also perpetuated the feeling that changing the training and education of architects will destroy this reputation and the teaching of pragmatic skills will inhibit the creativity of design work.

Despite Walker's criticisms there still persists in many schools of architecture and the built environment a highly subject-specific approach that in industry require participation, collaboration and integration of disciplines. The formal bodies representing the key built environment disciplines such as the Royal Institute of British Architects (RIBA) Chartered Institute of Building Services Engineers (CIBSE) and the Royal Institution of Chartered Surveyors (RICS) have in recent years tackled the issue of integrating sustainability into courses they accredit which serves to bring about a degree of educational change with potential for improvements in working practices.

The Egan Review *Skills for Sustainable Communities* highlights that a lot of work is yet to be done in establishing an educational ethos amongst those who are responsible for training built environment professionals in relation to the value and benefits of inter-disciplinary working and collaboration because 'it is very difficult to do so' [9]. The lesson can be hard to get across to students and even university staff. Frederick [224] and Cowan [225] assert that many built environment professionals often feel, and may well be taught from early on, that they alone have the ability to help shape a better world, and other, perhaps less creative or less rational professions, represent obstacles to the ability of their individual profession to achieve its own ambitions and aspirations.

Arguably, sustainability requires a higher order of interdisciplinary working because of the use of innovative materials, methods and technologies. The sustainable agenda in industry tends to move faster than in formal undergraduate education. There is a time lag because of the filtering of industrial practice into set curricula which takes time to reconfigure and accredit and educationalists tend to represent their own period of training and practice from the past. Of course, universities can lead industry when academic research is fostered to advance education for more collaborative, sustainable and integrative practices.

Numerous universities challenge their communities to save energy, water and reduce waste, challenging students in halls of residence to beat each other on savings. One such initiative is 'Green Impact,' an environmental accreditation scheme that encourages pro-environmental behaviours. It empowers sustainability champions within their workplace, helping them gain recognition for their environmental efforts whilst playing on the competitive spirit of staff working in teams. It provides people and their departments with a tangible framework for improving their environmental performance, breaking down complex environmental issues into more manageable tasks.

In the field of successful organisational change for sustainable education Hunting and Tilbury [226] offer a six-point model:

1. Adopt a clear, shared vision for the future.
2. Build teams, not just champions.
3. Use critical thinking and reflection.
4. Go beyond stakeholder engagement.
5. Adopt a systematic approach.
6. Move beyond expecting a linear path to change.

3.7 Chapter Summary

It is argued that all stakeholders throughout the design, construction, operation and use of buildings, in their attitudes and behaviours, should ideally engage with the sustainability agenda, particularly if they are to deliver on national and global requirements to limit carbon emissions. However, this is not often the case and buildings present a complex set of processes in relation to human behaviour where sustainability is often seen as an inconvenient additional layer of complexity.

Organisational arrangements and cultural beliefs tend to perpetuate the status quo in the built environment professions and limit the adoption of sustainable building practices. Overcoming these obstacles will require alterations in organisations beyond new mission statements and financial analyses. These changes must integrate environmental concerns into the existing routines by which buildings are constructed, recasting them in ways that are mutually beneficial to the objectives of individuals, organisations and the sustainability of the ecosystem on which they depend.

How we use building systems, services and appliances makes a significant difference in pro-environmental behaviour and it can simply be a question of breaking unsustainable habits and misperceptions that resources are abundant and our individual behaviour has little impact. Such ways of behaving are shaped and sustained by the buildings we design, construct, operate and occupy. As pro-environmental behaviour becomes more of a social norm there is a growing body of evidence that people do forego selfish benefits and assume personal responsibility and endure some personal sacrifice.

One key to encouraging pro-environmental behaviour is to remove social and economic barriers. People consciously and sub consciously tend towards congruency in their environments, therefore in a sustainable building where operation, services, technologies and materials are inherently sustainable, it is argued that sustainable and ecological behaviour is more likely to result.

Environmental psychologists and educators can be involved in the planning and design stages of construction and can even have a special role in evaluating completed buildings, to determine whether the goals envisaged in the planning and design stages are fulfilled in the completed building.

It has been identified that architects and educators need to understand each other and develop a shared vocabulary encompassing educational theories, developmental requirements, aesthetic theory, education for sustainable development and practical issues of designing schools. Future generations will reap the rewards of this integrated approach when they are able to occupy and use spaces designed expressly to stimulate their natural curiosity, where architecture is not a vacuous space but a learning tool to encourage and instil sustainable behaviour.

The following chapter draws together the evidence gathered so far and highlights opportunities for sustainable behaviour change during the building lifecycle as well as barriers, risks resistance and opposition to sustainable construction in general.

Chapter 4
Building Lifecycles and Behavioural Change

4.1 Introduction

Whilst the preceding chapters have focussed on the theoretical and philosophical issues surrounding sustainable buildings and behavioural change, this chapter investigates practical opportunities and applications throughout the lifecycle of buildings for changing the behaviour of individuals, organisations and institutions for greater environmental, economic and social sustainability.

Great advancements have been made in terms of the design and construction of low environmental impact buildings through technical innovation and an understanding of sustainable design principles. However, no matter how good the environmental credentials of sustainable buildings are, there are often ignored or unconsidered consequences of the environmental impact of human behaviour as a result of design decisions and construction practices on the performance of the building, often resulting in large discrepancies between envisaged and actual social, environmental and economic outcomes. Pro-environmental behaviour change is not currently one of the key behaviours envisaged by all but the most sustainability-minded design teams.

Decisions made throughout the whole life cycle of a building will affect energy use, CO_2 production, pollution levels, waste and the use of finite materials and resources, therefore it is argued that human behaviour needs to be considered throughout the lifecycle of a building, and sustainability should be considered from the outset to encourage, enable and facilitate sustainable behaviour of all stakeholders. Buildings and landscapes have a strong influence on our perception of the environment around us. A sustainable building should explicitly and implicitly influence the pro-environmental behaviour of its designers, builders, occupants and visitors.

Projects need to be controlled or influenced at every stage from inception to completion and operation. Opportunities during the project cycle for influencing environmental and sustainability issues are summarised below in Fig. 4.1.

J. L. Clarke, *Sustainable Buildings for Sustainable Behaviour: Theory and Practice*, Green Energy and Technology, https://doi.org/10.1007/978-3-031-98905-6_4

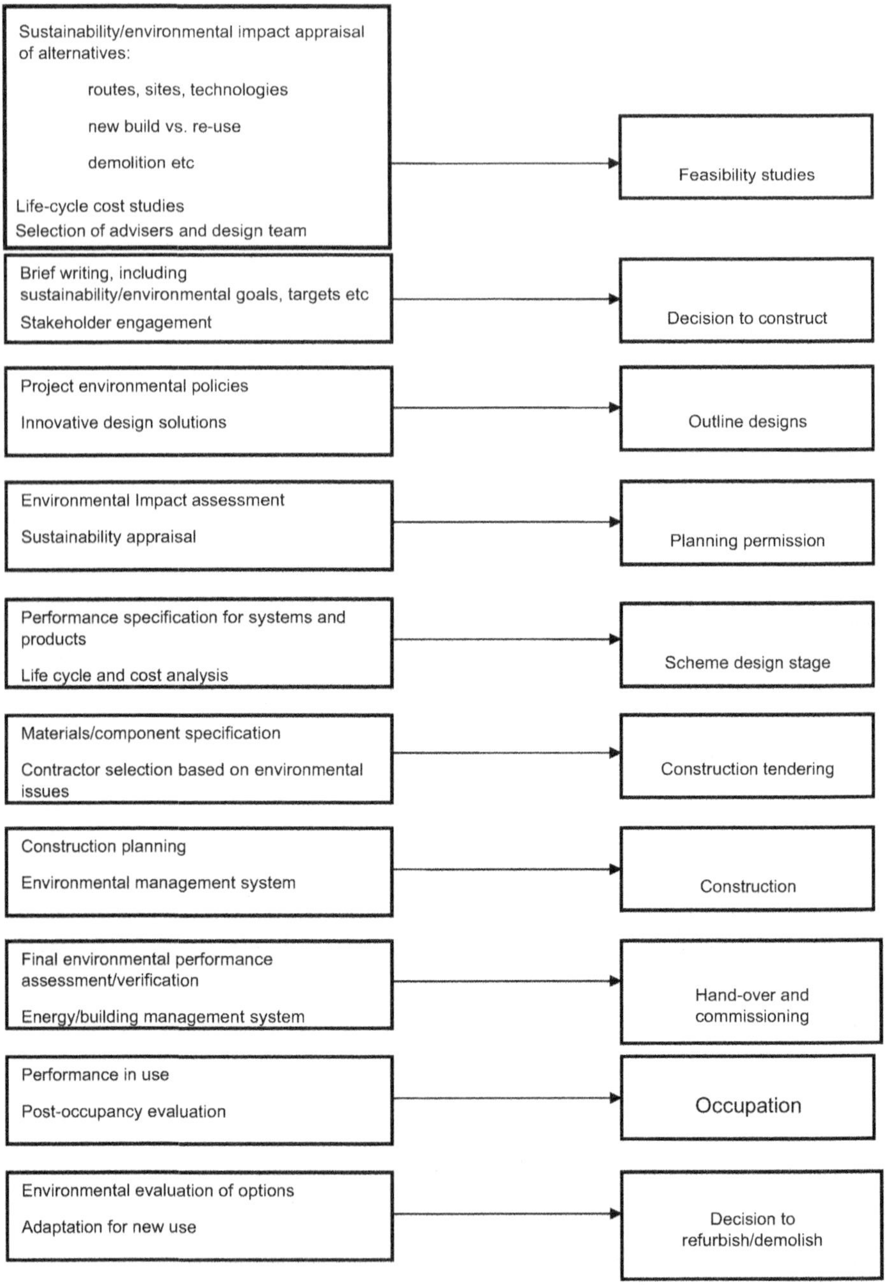

Fig. 4.1 Opportunities during the project cycle for influencing environmental and sustainability issues. *Source* CIRIA [227], p. 57

4.2 Organisations, Sustainable Behaviour and Buildings

The previous chapter focussed on how individuals behave in relation to issues of sustainability and the underlying psychological, pedagogical and emotional factors affecting both barriers and drivers to effecting behavioural change in relation to the built environment.

Professor Shove [180] suggests too much responsibility is placed on the individual, whether that is the architect or the eventual user of a building, which deflects attention away from institutions that shape options and opportunities. A survey by the Carbon Trust [228] indicates that over 75% of UK employees want to work for organisations with active policies on cutting greenhouse gas emissions. The same survey reveals that <20% of organisations actually run an energy saving programme.

Research by Allison [229] suggests that information available to individuals regarding the viability of green building options becomes a reflection of subjective organisational goals, routines and culture as much as objective facts. Professor David Orr asserts that:

> While we often speak of smart people, we seldom refer to smart organizations. Organizations of all kinds have great difficulty in learning so that their collective behavior often falls short of any reasonable standard of intelligence. Smart people working in dumb organizations create shoddy products, self-perpetuating conflicts, ecological ruin, boredom, inequality (and) idiotic doctrines. We do things in organizations that no one would do, or admit to doing, as individuals.
>
> (Orr [230], p.159).

Hoffman and Henn [11] identify organisational barriers to sustainable construction as being due to internal structure and interaction, language and terminology, rewards and organisational resistance and are discussed below.

Internal Structure and Interaction

The design and construction of a building creates a distinctive form of organisation in which a temporary culture becomes set, one which includes the roles, decision rules and power balances among each of the team members. Power and influence in the team can be a critical factor, often compromising the overall sustainability of the project as new environmental technologies and practices are introduced. Often when sustainability requirements are 'added' to a standard construction project, the roles and relationships become reconfigured into a form that is outside the standard operating procedure. This will invite resistance.

As discussed previously, structural relationships within the design and construction team are traditionally linear; client-architect-engineer-contactor-subcontractor, which tends not to promote integrated working or the tight integration of sustainable systems (water, heating, power) needed in a high performing sustainable building. In a sustainable building, the team must engage early on and in a more integrated and collaborative fashion and incorporate a wide body of knowledge from a variety of disciplines. Team members are challenged to discuss and adjust parameters that are traditionally made in isolation.

Language and Terminology

Many of the new sustainable technologies, processes, materials and standards involved in sustainable buildings comprise of new terms that may not be understood or may be misinterpreted and can therefore cause resistance to adoption. Also, the language of sustainability challenges conventional terminology requiring a new knowledge base. Kempton et al. [231] propose this new terminology can identify participants in the team who aren't yet embedded in the sustainable construction industry. Lack of environmental literacy makes the link between energy conservation and climate change more difficult for people to understand and creates a reduced sense of urgency or motivation for addressing environmental issues, much less to develop sustainable building practices.

Organisational Resistance

Peter Senge believes the core learning dilemma that confronts organisations is that 'we learn best from experience but we never directly experience the consequences of many of our decisions' [14], p. 159. Organisations tend to resist change [232]. People within them generally prefer long term certainty of structures and routines that have been historically and habitually in place for a long time and resist the process of changing them. Habitual routines can take form in 'taken-for-granted' design practices, construction methods or operational routines. Typically, the costs of learning new forms of sustainable design and construction practices are not charged to the client. With fixed resources, architects and contractors must currently invest in this learning process potentially to the cost of other critical activities. Arguably, the costs of such training and education could be spread more evenly throughout the industry and by government agencies.

Fear of the unfamiliar and policy uncertainty can also drive organisational resistance, particularly when the consequences of change cannot be predicted and can put individuals off embracing sustainable practices in the future. Psychological blocks can prejudice managers away from certain actions or responses to demands for change. This may deny a developer, architect or engineer any opportunity to consider longer term gains. Cost is an important issue for the client and the design and construction industry and when financial decisions are made on whether to invest in sustainable features and funds are limited, often the sustainable option is the first to be rejected under the justification of value engineering. There needs to be full consideration of capital costs versus lifecycle costs and payback periods which are rarely factored in to project costs and are often not seen as a priority for both architects and contractors. It is increasingly difficult to quantify areas such as projected future energy costs.

Consideration for the adoption of innovative sustainable technologies requires more time and effort. New technologies must be identified, integrated and tested as the technologies themselves develop and improve. Time pressures can prevent full investigation and solving of environmental issues in the production of sustainable buildings. Mintzberg [233] highlights that sustainable design and construction can challenge established authority within organisations which can result in interdisciplinary rivalry or organisational resistance. Chamberlain [234] questioned whether the addition of a new skill set falls to the architect, contractor, engineer or a new green or integrative design consultant? Existing professionals in building design and construction may resist these changes to protect their professional status. Arguably, it is the responsibility of the entire built environment industry to upgrade sustainability skill sets, just as is it for health and safety and other statutory obligations.

Howard-Grenville and Hoffman [235] conclude that adoption of changes in practice is easier if presented as a positive and attractive option rather than as an issue of sacrifice. Some professionals are put off by the phrases 'green' or 'sustainable' buildings and are much more engaged by terms such as 'smart,' 'intelligent' or 'high performance' buildings. The WISE building case study, investigated in the following chapters, has one of its main design aspirations as having a clean, modernist and professional looking building, deliberately avoiding references to its 'green' credentials in order to appeal to mainstream organisations for holding exhibitions and conferences, with the strategic aim that the sustainable features will impact in a more subtle and subliminal way and evidence is presented to support this.

4.3 Changing Behaviour Through Institutional Change

Policy and technological innovations can strongly influence individuals and organisations in their consumptive behaviour and arguably the greatest influencing factors in terms of social, economic and environmental behaviour change for new buildings are the building codes and regulations. They do however have limited control over existing buildings that require upgrading in terms of sustainability.

Barriers to sustainable construction can be perpetuated by rules, norms, and beliefs at the institutional level. Organisations are inescapably influenced by the external environment through both technical constraints such as raw materials, labour and energy, and through social influences embodied in rules, laws, industry standards, established best practices and conventional wisdom—collectively referred to as institutions by Scott and Meyer [236]. Institutions present cultural and contextual constraints that alter individual and organisational perspectives. Hoffman and Henn [11] offer three categories under which they analyse the influence of institutions on the adoption of sustainable construction practices: regulative, normative, and cognitive.

Regulative Considerations

Regulative or legal aspects of institutions are based on coercive or legal sanctions to which organisations accede for reasons of expedience. Easterbrook [237] asserts that environmental standards have generally been beneficial and have produced positive results. Problems and costs can arise from a regulatory approach to environmental problems. Tenbrunsel et al. [238] propose that legal standards become an independent force taking on a life of their own, leaving rationality, innovativeness and societal interests behind. They suggest that suboptimal outcomes can result from an adherence to standards due to a tendency for standards to direct attention towards the law itself and away from the purpose behind the law, known in behavioural psychology as the 'mis-directed attention effect.'

For example, energy regulations for new construction may not be updated for several years, and efforts to do so become protracted political battles that result in compromise that presents a significant barrier in promoting sustainable design. Another example, prescriptive U-values, penalise high thermal mass walls that contribute to passive solar design. Once standards are written, decision makers within organisations often become constrained by rigid rules that preclude the search for creative solutions to complex environmental problems. At times, these standards can explicitly restrict environmentally optimal solutions.

Tenbrunsel et al. [238] suggest a psychological explanation for the misdirected attention effect, namely, that standard-based systems can change the incentive system for individuals and promote self-interested behaviour that conflicts with overarching societal interests. Hoffman and Henn [11] found that unintentional actions can result from individuals 'just following the rules' or having a 'no law against it mentality.' However, intentional actions include trying to 'beat the system.' Some have suggested that the Building Research Establishment Environmental Assessment Method (BREEAM) system itself has suffered from the misdirected attention effect. Critics charge that BREEAM has become a 'point chasing' or 'tick-box' exercise with participants losing sight of the objectives of sustainable buildings, to minimise the environmental impact of the built environment, and instead focus on achieving the highest rating with the least effort.

All construction projects need to comply with legislation. Legal requirements define the minimum environmental performance criteria that a project should meet. Currently, many environmental and sustainability targets are not addressed or enforced by legislation or regulation. An early decision to be taken by the client and design team is how far above the minimum to aim and the degree to which these might be achieved, for example, very low energy, zero or even negative (i.e. contributing to the national grid).

Normative Considerations

Normative (or social) aspects of institutions are morally or ethically grounded, and organisations will comply with them based on social obligation. These take the form of rules of thumb, accepted economic indicators, standard operating procedures, occupational standards, educational curricula, and membership requirements, which emerge through universities, professional training institutions and trade associations. The building industry is highly structured with standards set by a wide variety of organisations. These organisations specify detailed parameters by which products must be made, buildings must be built, and future professionals must be trained.

Hoffman and Henn [11] assert that the integration of environmental concerns into these standard setting bodies moves very slowly. For example, structural engineers feel secure specifying concrete meeting the same standards they have safely used for years even if new and more sustainable (but less familiar) materials meet the same or adequate performance criteria, such as limecrete or rammed earth. Other market incentive structures yield similarly environmentally inappropriate behaviours. Architects and engineers are often compensated on a percentage of the cost of construction, essentially penalising them for eliminating costly practices. Landlords have no incentive to improve the energy efficiency of their buildings because tenants pay for energy costs.

Lovins and Lovins [239] focus their research on energy pricing which allows regulated utilities to increase profits based on increased energy use and, in effect, penalises them for encouraging reduced energy consumption. As a result, shareholders and customers have opposite goals with increased energy use as the end result. Banks are unwilling to provide financing for certain environmentally sound technologies fearing that they are unproven or believing that they are unnecessary. For example, lenders will not generally provide financing for renewable energy systems if more traditional and secure grid-connected power is available. However, recent policy changes related to certain renewable energy technologies in the form of subsidies such as the Feed in Tariff (FITs) and the Renewable Heat Incentive (RHIs) and the introduction of the Green Investment Bank are proving to have a significant impact on the adoption of renewable energy technologies but are sensitive to policy uncertainties. Building inspectors resist innovations integral to sustainable buildings because they are new, unproven and in many cases uncertified.

Cognitive Considerations.

Cognitive (or cultural) aspects of institutions are the accepted beliefs and common perceptions of behaviour which organisations will abide by without conscious thought and are taken for granted and remain unquestioned, as described by Zucker [240]. These are pervasive, powerful and resistant to change. The construction of the built environment is a practice so old that it is not surprising to find innumerable unquestioned biases and assumptions.

Institutional structures permeate our beliefs and thoughts. Coupled with individual and organisational biases, they form systemic aspects of our society's resistance to the adoption of sustainable building practices. Integrating a concern for social, economic and environmental sustainability in the built environment will require adjustments in the overall system in which buildings are designed, constructed, operated and used. Changing practices will require addressing some underlying beliefs about why and how we build.

Strategies for overcoming the social and psychological obstacles to the adoption of sustainable buildings as normal and mainstream practice fall into two categories according, to Hoffman and Henn [11]: (i) as an entrepreneurial opportunity or (ii) as an obstacle to be overcome. In both cases, strategies cannot be targeted strictly at the individual, organisational or institutional levels. Successful strategies create behavioural change across all three levels of analysis.

Further Opportunities for Behavioural Change

Sustainable methods, materials and technologies throughout the lifecycle of a building can offer opportunities that are more economic and efficient than existing practices. If an organisation begins to tackle sustainability issues for the first time it can begin to make significant savings relatively easily, e.g. waste reduction and energy efficiency. Some firms are more likely to capitalise on opportunities than others. An organisation's proximity to final consumers makes the firm more likely to participate in environmentally friendly behaviour in response to green consumer demand. Moon and De Leon state that 'these organisations are more visible to consumers and susceptible to green publicity because product sales are largely associated *with* the publicity' [241], p. 484.

Berkhout and Rowlands [242] noted that firms which publish environmental metrics and espouse environmentalism as an organisational value are more likely to undertake sustainable practices, e.g. purchase green electricity. Building sustainably can create a comparative advantage, even suggesting product-price premiums because of their green reputational credentials. If social and psychological barriers are inhibiting other organisations from realising such an advantage, entrepreneurs will step forward to fill the void. There are many successful architectural practices, construction firms and developers who have recognised this, often allied with genuine philosophical and ethical considerations of sustainable building practices. Some new consulting firms provide an audit and installation service of energy conservation technologies asking only to be paid based on a percentage of energy savings.

Some insurance companies, such as the US Fireman's Fund, recognise that insuring buildings requires specialist knowledge of new building systems and offer policies that assure damaged elements of buildings are replaced with green products or whole buildings are replaced according to recognised environmental standards offering a premium discount due to identified lower risk factors of sustainable buildings.

Some banks and other lenders are beginning to offer energy efficient or green loans, such as the newly created Green Investment Bank, recognising that a building owner with lower utility bills can afford a higher mortgage payment. Certification programmes (e.g. BREEAM) fill a market void by making green labelling clear, transparent and objectively defined. Such programmes have been central in overcoming the social resistance to sustainable construction.

Agents for Change

For fundamental behaviour change to occur it is important to identify and target demographic adopters, prime movers who are more likely to accept new technologies and to take risks on sustainable buildings. U.S. research by the Natural Marketing Institute [243] shows that certain elements of society are more aligned with environmental values than others. Marketing professionals dub this demographic group LOHAS, signifying Lifestyles of Health and Sustainability, based on gender, age, education and income, urban versus rural and political affiliation.

This research shows that women are generally more environmentally aware than men. Young people tend to be the most environmentally aware group with the second most aware group from 36 to 45. Environmentally driven choices tend to increase with both education and income levels. People in urban areas tend to use environmental considerations in their purchasing decisions more than people in rural areas and those who support 'left leaning' political parties believe in anthropocentric climate change.

In a newspaper article by Monbiot [177] data from an ICM survey showed people in social classes D and E (working class to those at the lowest levels of subsistence) who thought the government should prioritise the environment over the economy was higher at 56% than the proportion in classes A and B (upper middle to middle class) at 47%. This indicates that rather counter-intuitively, environmentalism is not just an issue for the middle-classes. The social grades referred to are taken from the National Readership Survey, a commonly used market research standard [177], p. 29.

4.4 Pre-construction and Sustainable Behaviour

Planning and Pre-design

At pre-design stage the means of integrating sustainability and environmental features must be identified and how they may be achieved by adopting appropriate plans and systems of sustainable management and procurement processes. The sustainability policies adopted need to translate into a consistent and coherent set of policies, strategies, procedures and process for delivering sustainable construction. These strategies and plans will be the foundation for managing sustainability issues throughout the whole construction process and building sustainability principles and practice into construction to deliver all round social, economic and environmental value to all stakeholders.

The client must be fully open to the opportunities that sustainability can offer and address these issues based on a good understanding and sound advice very early on. Most difficult of all is assessing the value to a particular stakeholder of achieving performance levels above minimum standards and requirements. Generally, the input needs to come from environmental and sustainability specialists who can advise the client and provide a co-ordinating function for the project with the necessary overview and breadth of understanding of the environmental, social and economic issues involved, and to co-ordinate this across the entire design and construction team.

The development of new procurement strategies is an approach increasingly being adopted by construction clients trying to achieve improvements in quality and value and to allocate risk where it can best be managed. While the key factors in the delivery of sustainable construction are the skills, experience and knowledge of the client and project team, there is a growing body of evidence to show that adopting the appropriate procurement strategy has a significant impact from a sustainability perspective.

The selection of a procurement route involves a systematic assessment of alternative forms of contract available to the construction industry client, compared in terms of time, quality, cost and sustainability. Each route places different demands, risk allocation and responsibilities on all parties involved. Construction partnering agreements can encourage clear guidance from the client, effective management of the project and good understanding and sharing of environmental goals among the design and construction team. Experience with similar projects and a variety of procurement routes is also a critical factor.

The Egan Review [9] strongly recommends the wider use of partnering to encourage organisations to work together to improve performance through agreeing mutual objectives. Partnering can help to resolve disputes, assist in continual improvement, promote performance measurement, lead to a sharing of both positive and negative experiences and assist in the recognition and allocation of risk—all essential preconditions of achieving sustainable construction.

A critical part of any procurement strategy directed towards delivering sustainable construction must be the selection of consultants and contractors with tried and tested levels of commitment and experience. This is facilitated by the trend in construction procurement away from fee competition, in the appointment of consultants, towards a selection process based on the balance of quality and price, which followed the influential reports on the construction industry by Egan [9] and Latham [244]. Regardless of the procurement method chosen, project partnering can achieve savings of 2–10% in the cost of construction [245].

Value Management

Sustainable construction can often deliver added value to projects, and value management can be a fundamental tool for ensuring this is maximised. The aim of value management is to ensure the client's objectives are fully articulated and understood, and to meet these in the most efficient way. The process involves listing the client's needs and prioritising them, under the guidance of an experienced facilitator. A value management workshop can help the design and construction team meet the client's needs in a non-confrontational environment. Value management differs from value engineering in aiming to achieve best value rather than lowest cost. It can help meet common objectives through reduced project time via project simplification, increased site productivity via better design, allowing easier assembly, maintenance and demolition, increased certainty of completion on time and to budget and reduced construction and life-cycle costs by having less complicated installations. Value management takes account of factors such as:

- the balance between capital and operational costs
- innovative design alternatives
- the application of risk management
- environmental impact and sustainability
- life-cycle replacement costs
- buildability
- whole-life issues, including adaptability and eventual demolition

Benchmarking

In order to remain competitive, leading organisations need to compare their own products, services and business practices with others, inside or outside their own industry, seeking to identify and implement best practice. Benchmarking is a process for achieving continual improvement by setting targets and measuring performance relative to the exemplary practices of others. The idea of a benchmark is to provide people who are working on a particular project with an indication of the environmental performance achieved by similar projects. While not necessarily representing targets that must be achieved, it allows a project team to choose whether it is going to aim to be typical or much better than is usual in certain aspects of environmental performance. A benchmark may refer to relevant products, processes and performance.

The Display Energy Certificate (DEC) calculates a non-domestic building's emissions per square foot and then rates that building against others of a similar type, the higher the rating the more energy efficient the building is considered to be. This can be misleading, as buildings with a low occupancy level could have a higher DEC rating simply because they use less energy overall, even though that usage may be more wasteful than a high occupancy level building. A number of benchmarks are being developed to address these issues, including the Low Carbon Workplace (LCW) standard which measures CO_2 emissions per person per year. These enhanced benchmarks give a more rounded view of carbon emissions for buildings.

As already identified, the performance of buildings in use are often found to be at variance with their designer's expectations, sometimes because of the unreliability of design methods, but more often because the buildings have not been built as the designers specified, and most likely because they are operated incorrectly by the building occupiers. These sources of data can provide design teams with suitable benchmarks.

Leading companies in many sectors of business, including an increasing number of UK construction companies, are already measuring and benchmarking their sustainability impacts. These companies recognise that a reorientation towards sustainable development is essential for ensuring their long-term viability.

Whole-Life Cost Studies

Whole-life costing is gaining prominence for building projects. The owners of buildings are beginning to look beyond a short-term view based on capital costs and are demanding evidence of what the long-term cost of ownership will be. Methods of procurement, such as private finance initiatives and public private partnerships, focus attention on the long-term assessment of costs as a basis for investment decisions. Whole-life costing allows purchasing and investment options to be evaluated more effectively by taking into account the impact of initial costs, operating costs, management costs, maintenance costs, disposal costs, opportunity costs and other costs such as insurance. This permits the optimisation of construction spending and timing and allows for better long-term management of assets. Typical benefits of whole-life costing include.

Investment

- provides a balance between capital and operational budgets
- assesses alternative courses of action
- identifies cost effective designs

Future Proofing

- investigates the sustainable development of limited natural resources
- evaluates the burden being placed on future generations to maintain buildings and infrastructure
- enables the financing of future commitments

Management

- addresses operating cost cash flows over the life of the project
- enables continuous cost management throughout the stages of design, construction and occupation
- facilitates data collection and feedback from existing projects
- provides information for financial planning and analysis of future expenditure
- monitors the success of design options against estimated targets and highlights possible areas of improvement

Whole-life costing should be carried out by the client at various stages of the decision-making process, with appropriate refinements at each stage to reduce the uncertainty inherent in construction and operational costs.

While the benefits of whole-life costing are becoming more widely understood, barriers still stand in the way of a more widespread acceptance of the approach. The effectiveness of whole-life costing crucially depends on the capacity to predict accurately both operating and capital costs. This is not as easy as is sometimes supposed since few organisations have reliable data about operation and maintenance costs and the long-term performance of components and materials. Long-term owners of buildings in sectors such as social housing, education establishments, hospitals and hotels have an advantage here, but even their historical data can be inappropriate for the type of sustainable buildings that are being constructed today.

The benefits of whole-life costing also require the possibility in the project structure of transferring the financial benefits of reduced operating costs into the capital budget for a project. This is often, still, surprisingly rare. All projects should be set up to ensure that this is possible. Typical examples include: design, specification and construction to avoid maintenance and optimise durability and better insulation in the building envelope in order to reduce long-term heating costs.

Different organisations will have different expectations of building life. Not all client bodies want or need their buildings to last 60 or 100 years. At the end of a relatively short building life, residual value and demolition costs may be significant factors. The arbitrary figure of a 60-year lifespan has its origins in financial modelling, the time over which financial assets are written down and pays little attention to social or environmental considerations.

Despite the general acceptance that buildings should last up to 60 years, there is growing evidence, highlighted by Oliver [246] that a minimum requirement for the lifespan of a building, making longevity a key objective, would have a huge impact on the way buildings are designed. The focus of the designers would then be on the performance of the building in use rather than on the embodied energy of the original structure. The focus should be on building sound, energy-efficient structures that will last several life spans, with the energy efficiency built into the fabric. Proposals have been made to adopt a longevity calculator to define a minimum acceptable lifespan for the main structural elements of the building fabric. There is also the question of value—buildings that are valued are more likely to last longer and be well-maintained.

A long-term problem with sustainable buildings exists in terms of their becoming redundant in the light of new technologies, methods and processes. A state-of-the-art green building to suit client's business needs and public image can soon appear outdated and even wasteful by new contemporary standards. At the current pace of innovation in the sustainable built environment sector it is likely that a building would be lucky to keep its position for more than about 5 years, far less time than the likely period for which an owner-occupier might want to use it.

Future-proofing a building makes it possible to anticipate changes and trends so that a building becomes amenable to the sort of changes that would be needed to improve its performance and extend its usability. New sources of energy could be used, new IT and other building services could be installed, a building facade could be replaced by one with better thermal performance, the internal structure and layout could be altered to suit new ways of using the internal spaces. The difficulty is being accurate with such predictions in a time when innovation in this field offers many untried and untested options.

Designing buildings for flexible patterns of use and constructing them in ways to allow minor adaptation such as re-routing services or installing more efficient solar technologies is gaining popularity in the sustainable construction field. If flexibility or adaptability is required it must be carefully specified in the design brief. Caution should be exercised as it is easy to be over-confident about trends and end up building in too much flexibility and adaptability, with consequent capital cost penalties.

Most buildings survive beyond their assumed lifespan and often outlive their originally-conceived function. Therefore, there is economic, social and environmental value in their reuse either through adaptations to small changes of use, for instance internal layout or major refurbishment, significantly altering the original building. If no plausible reuse can be found demolition is inevitable and the most sustainable use of the embedded resources of materials and components must be followed through the salvage, reclamation, reuse or reprocessing for reuse, rather than consigning them to landfill. In this case it is highly desirable to devise a waste management plan for demolition and materials segregation and salvage.

Design for deconstruction makes the process of demolition easier and the potential for reuse and recycling of materials and components more feasible. At design stage hazardous materials should be avoided, but if used they should be identifiable and easily separable to facilitate recycling. As with all issues that affect a building, they are best considered at the beginning of a project, for afterwards they are likely to be costly or impossible to introduce.

The Design Brief

The design brief provides the basis on which scheme and detailed design can begin. The identification of constraints that cannot be controlled by the designers is fundamental to the preparation of a complete and comprehensive design brief. The accountability of the design team for issues such as energy use, harmful emissions and waste generation needs to be clearly acknowledged.

While traditional design and construction focuses on cost, performance and quality, sustainable design and construction involves additional criteria—energy conservation and efficiency, minimisation of resource depletion and environmental degradation, and creation of a healthy pleasing and efficient built environment for living, working and leisure. To deliver sustainable construction these factors must be given high priority in the development of the brief.

An essential foundation of sustainable construction is the concept of eco-efficiency—producing more using fewer resources and causing less pollution. Sustainable construction means applying the concept of eco-efficiency across the full life cycle of building materials—from the acquisition of natural resources, transportation, product manufacture and construction, installation and use, to their ultimate salvage or reuse. Similarly important is the concept of 'dematerialisation'—the planned decline in material and energy intensity in industrial production, distribution and construction through design for disassembly. Cost-effective efficiency gains of more than 400% have been projected in some circumstances through the application of the principles of eco-efficiency and dematerialisation.

Specifying sustainability and environmental performance can vary from rather imprecise ideas of green building to very precise definitions. This is not only when the building is being used, but also during construction and the manufacture of all components that are used in its construction. A variety of means for describing and assessing environmental performance are available, including the use of benchmarking, key performance indicators (KPIs), life-cycle assessment (LCA) and environmental assessment methods such as BREEAM. By these means environmental goals can be incorporated into design briefs and contracts.

As the field of sustainable building advances, the standards that promote it must also evolve. Dominant standards must promote green building as a holistic and accurate process towards alleviating the impact of buildings on the environment. Standards should give flexibility for the construction team to focus on the right sustainable building technologies by developing site-specific and client-specific alternative strategies that achieve equal or greater environmental benefits at lower costs. Standards should also focus on the secondary effects of such regulatory programmes. Policies should begin to move away from a focus on direct, marginal and incremental mechanisms for bringing about individual change and should begin to stimulate both direct and indirect pressures for industry-wide change. Specifically, attempts must be made to change core business networks, such as financial markets, inspectors, insurance and consumer demands.

In addition to the mandatory standards enforced through the building regulations, there are a range of independent standards and this study focuses on one of the most well-established globally. BREEAM has developed over time to reflect the environmental lifecycle of a wide variety of building typologies, assessing environmental, social and economic impacts at each stage of design, construction, operation and use. BREEAM assessed buildings are awarded credits depending on their environmental performance and can achieve a rating of either unclassified, pass, good, very good, excellent or outstanding.

The choice to opt into having a BREEAM accredited building is complex, not least because of its perceived economic cost or value to a project and it can be an expensive endeavour. A recent Building Services Research and Information Association (BSRIA) report published in *Public Sector Sustainability* [247] found that less than half of respondents incurred significant extra costs on their latest BREEAM rated project, and four in ten said they did not incur extra charges compared to a non-BREAM project. For some respondents the additional costs of using BREEAM were seen as an investment for the future, with payback coming from reduced building running costs. The main reasons for adopting BREEAM certification were company policy/CSR (47%), planning requirement (33%), and regulatory or procurement reasons (16%). Economic benefits were low down the list.

BREEAM accreditation is encouraging organisations to reduce the lifecycle impact of their buildings, demonstrating their environmental commitments by having a credible label they can use and stimulate demand for sustainable buildings. The report also highlighted that a deciding factor influencing the cost to a project of adopting BREEAM was the point at which it was included in the design process. A number of respondents collectively acknowledged that the earlier the decision is made, the better the chance of keeping costs down.

Negative aspects reported were the perception that BREEAM does not consider operational costs and only 47% of those surveyed thought about operational costs versus capital build costs, showing that current levels awareness is low. When asked if respondents knew if the client recovered the cost of BREEAM 23% said they did not.

Two studies conducted by Surgenor and Butterss [106] and Jansz et al. [248] respectively have shown that the percentage increase in capital costs when undertaking BREEAM depends mainly on two factors; (i) target rating; when the target rating is higher the capital cost may be higher than less ambitious projects and (ii) location; a building well served by public transport and on a brownfield site would achieve a higher score. Moreover, the capital cost increase in achieving a high BREEAM rating varies according to the type of building assessed, as Jansz et al. [248] showed. The capital cost percentage related to achieving a BREEAM 'Excellent' rating ranges from 0% for simple naturally ventilated offices to approximately 7% for more complex buildings.

In a pedagogical context, the BREEAM Education category *'Management 10-development as a learning resource'* assigns additional credits for integrating sustainable features in buildings that have educational benefits for users, utilising the building and landscape as an educational learning and teaching resource.

Innovation credits are also available for sustainable procurement, responsible construction practices, reducing CO_2 emissions, use of low/zero carbon technologies, water consumption and flood prevention, life cycle impacts, responsible sourcing of materials and construction site waste management.

However, even when adopting this industry recognised standard there are many different approaches to achieving a sustainable building under the BREEAM guidelines with broad variability because of design limitations and/or low sustainable aspirations, wasteful construction practices, poor operation and inefficient use of buildings.

Several aspects of defining environmental performance make it difficult to be as prescriptive as structural, mechanical and electrical engineering:

- the number of environmental parameters is large and those related to human comfort are partly subjective
- the goal posts are moving—what was considered excellent environmental performance a decade ago, is only average now
- benchmarking data is usually well behind best practice at a particular time, and there are still many indicators of environmental practice for which no benchmarks are available
- best practice is far ahead of statutory requirements and is currently self-imposed by a particular client

It has been widely recognised that what is needed is an internationally agreed set of common metrics for measuring building environmental social and economic impacts. This will allow a means of measuring, reporting and verifying improvements in a consistent and comparable way. This is happening in a joint partnership between the United Nations Environment Programme Sustainable Building Climate Initiative (UNEP-SBCI), World Green Building Council (WGBC) and Sustainable Building Alliance (SBA), supported by the various rating tool providers.

4.5 The Design Phase and Sustainable Behaviour

Design for Behaviour Change

It is widely accepted that the greatest opportunity in the whole life cycle of buildings to impact sustainability is at design stage, both in terms of methods, materials and technologies used but also by setting behavioural goals encouraging, influencing or even manipulating the sustainable behaviour of individuals and organisations involved throughout the process, from inception to operation in use.

The explicit intention to change behaviour through design interventions has been named 'design with intent' and 'persuasive design' by a leading practitioner, Lockton [159]. Lockton describes this as design that is intended to influence or result in certain user behaviour. It is a strategic and reflective approach that recognises that designers not only influence products and experiences but can design to address specific target behaviours, such as sustainable behaviour. He has developed a toolkit of 101 cards that pose questions to designers rather than give prescriptive answers. He calls these 'provocations.' Examples are shown in Fig. 4.2.

Lockton explains that the key is pattern recognition of the problem and quick matching to possible moves to address it from multiple design-related disciplines during the early stages of the design process. Clune [8] undertook a 4-year action research study whilst teaching Design for Sustainability (DfS) and proposed that designers should engage in design for behavioural change to encourage positive actions that contribute to a sustainable society. He differentiates DfS from more traditional eco-design strategies that rely heavily on technical solutions. Clune's study highlights that design for behavioural change provides a process for intervening in unsustainable human activity, citing that 'design often holds unsustainable behaviours in place, making them difficult to overcome. The potential for design is to make our default actions more sustainable' [8], p. 68.

McKenzie-Mohr and Smith's [249] research on Community Based Social Marketing (CBSM) highlights the need to prompt people towards more sustainable behaviours by applying the psychological principles of behavioural change (as discussed in the preceding chapter) which show strong correlations between designers' skill sets and CBSM. Like the process of CBSM, design for behavioural change requires a clear directive from the designer as to what behaviours are to

Fig. 4.2 Examples of 'design with intent' provocation cards. *Source* Lockton [159]

be targeted. The more specific the targeted behaviours, the easier it is to tailor an intervention.

Hidden, positive actions that promote sustainability need to be made visible and desirable in-built form as social norms that can be readily followed. Creating norms through visualising possible futures is a strategy that has a history of use in industrial design. The visions presented by early industrial designers of possible futures conditioned our normality and paved the way from those visions to become what is now our carbon-based and unsustainable reality.

> The issue is not only what design does to people but what people do with design
>
> (Walker [250], p. 183)

Building design can influence user behaviour to reduce the social, economic and environmental impact of buildings during design, construction and operation. Practitioners who combine behavioural science and design have identified various interventions that have an effect on behaviour change in relation to product design and many of these can be transposed to built environment practices.

Building design-led interventions include building materials, technologies or techniques which attempt to modify behaviour to reduce the environmental impacts of building design, construction and use to facilitate all stakeholders in order to adopt ecological behaviour through the prescriptions and/or constraints of use embedded within and throughout the building, these interventions should ensure and maintain behavioural change. Such interventions might include making the structure or operational systems transparent, providing feedback on building performance, encouraging interaction with sustainable materials or simply encouraging user-control of their environment by incorporating openable windows.

Specific behavioural goals must be identified and set for the design plan, although a behavioural goal is just one of several types of goals that must be considered, including structural, financial, aesthetic and others. Those target behaviours serve as the starting point in the search for sustainable features that might support or facilitate them. Then the design or plan elements that are identified as likely to enhance the behavioural targets must be incorporated into a complete design or plan that specifies materials, layout, technologies, pedagogical considerations, sizes of spaces and relationships among those elements. This may require the resolution of serious conflicts between sustainable behaviour-related goals and the interests of other parties to the design or planning process.

Sustainable design needs new ways of thinking and working, and people prepared to take the time for this to happen.

Addis and Talbot [251] suggest seven key aspects to managing sustainability and environmental issues during design:

- Ensure environmental performance and sustainability issues are on the agenda of appropriate project and design team meetings.
- Ensure the client's environmental or sustainable aspirations and policies are covered in the project design briefs in ways that can provide the design team with the appropriate direction.

- Define appropriate sustainability and environmental performance criteria and targets.
- Ensure the design team has appropriate knowledge and skills to undertake the environmental and sustainability aspects of the design or, if not, seek appropriate specialist advice.
- Arrange regular assessments of the project's predicted environmental/ sustainability performance in order to monitor progress towards the targets stated in the brief, e.g. BREEAM rating.
- Ensure adequate interaction between members of the design team in relation to preparation of environmental impact and intended sustainable behaviour assessments. This will instil a thorough understanding of the impact of their decisions and actions.
- Provide reports on progress during design to the client, project manager and design team including a full set of environmental performance and sustainable behaviour targets, for both the construction and operational stages of the project [251], p. 145.

Integrative Design

Integrated design characterises what architects do when they incorporate the energy, site, and climatic, formal, construction, programmatic, regulatory, economic, and social aspects of a project as primary parameters for design. Reed [252] contrasts the term integrated design with integrative design, arguing that the former implies something that is past and completed and the latter which suggests an evolving, rather than a fixed process. The term integrative design is adopted in this study as an exemplar of best practice.

The integrative design process is a collaborative approach to building design which places a strong emphasis on cross-team integration throughout the development process in pursuit of a 'whole building' holistic design. In conventional building design a project develops through a strict and rigid chain of milestones and hand-offs, e.g. owner requirements to architect, architects' concept to structural engineer, structural design to mechanical engineer etc. This conventional approach means that key members of the design team are often excluded from the critical planning stage, and with the lack of their expert knowledge and insight the project can progress in the wrong direction for a significant period of time before serious underlying problems are identified. This leads to inefficiency, higher capital costs, time delays and over-sized services, and so on.

With an integrative design process all key members of the multi-disciplinary design team are included at the very beginning of the planning stage, from the initial conception of the building itself. In this way all major design decisions can be carefully considered in relation to other disciplines from the outset. This avoids abortive work resulting from single-minded decisions and increases overall project efficiency. Interdisciplinary procedures raise awareness of other disciplines concerns and add to the holistic design approach. *Intra*disciplinary working, where one person works within the discipline of another, is also a highly effective way of understanding different roles within the design team with the opportunity for changing knowledge, attitudes and behaviours.

An integrative design process—starting from a shared vision—can address many of the problems associated with current mono-disciplinary practices and helps to ensure that sustainability objectives are managed and met in a comprehensive and co-ordinated way. However, in a recent article by Flood [253] it was argued that the property and building industries are fragmented. He highlights that employers often separate development teams from asset management teams, while contractors are appointed separately from facilities managers. This does not help with joined-up thinking in terms of building design, procurement and life cycle matters.

Assembling a design team with clear commitment to the concept of multi-disciplinary working—shared by the client and contractor—is an essential pre-requisite. Spending time up front can save time and money later in the project when difficulties caused by lack of co-ordination or integration are avoided. Such understanding is leading to a fundamental review of the design process and the role of design professionals within it.

Clearly, the design of a building has long-lasting implications on its surroundings, infrastructure, constructors, occupiers, operators and a whole myriad of people engaged in its operation and use. A particularly illuminating social theory approach to viewing architecture as an integrated process is actor-network theory (ANT) developed by science and technology studies scholars Michel Callon and Bruno Latour, the sociologist John Law, and others.

As presented by Fallan [220] architecture can be seen as a process and socio-technical network of complex interactions between people and artefacts. Typically, architecture critics see the conclusion of the work at its most pure and uncontaminated at a single point in time. This is how architecture is usually portrayed in architectural journals and books, without seeing construction sites or buildings in use, devoid of the human aspect, despite the rhetoric of architects speaking of places for people and social interaction. Fallan calls this the 'stasis of perfection' [220], p. 88 which ignores the post-occupancy phase of a building when users have been shown to have a powerful impact on the use, form and performance of a building. So to understand sustainable architecture one should examine what its users make of it. This justifies the use of the case study approach, not just focussing on design and construction but also the operation, performance and use of the building through interviews and questionnaires.

Gieryn [254] counters the classical dualist perspective of architecture, with human actors being separate from structure, reinforcing the idea that architecture is socially formed and that architecture also informs social behaviour, whilst in turn being continuously transformed by social attitudes. This supports the main hypothesis that sustainable buildings and sustainable behaviour have a reciprocal relationship.

This was recognised by Winston Churchill in a well-known quote after the bombing of the House of Commons in 1943 'we shape our buildings, and afterwards our buildings shape us' [254], p. 35.

Gieryn goes on to assert that the ANT method is strongest when one aspect of a building's design is analysed and demonstrates how different concerns from different actors are consolidated and translated into a particular solution. This strongly supports the application of ANT to the design, construction, operation and use of sustainable buildings and to view architecture as a collaborative process requiring the integration of many actors, particularly as buildings and systems become increasingly focussed on performance and sustainability.

Fallan [220] highlights perhaps the most provocative and controversial aspect of ANT as the concept of non-human actors and their influence on networks. Artefacts as actants relate well to architecture, as architects have a special sense of the properties of materials and material objects and an acute awareness of, and interest, in the impact that our material surroundings have on us. Materials possess distinct characteristics, qualities and capacities. This is particularly true of sustainable materials which are key actants in the sustainable building network with implications on health and well-being, biomimcry, vernacular architecture, locality and embodied energy.

According to ANT a building is a technological artefact constructed by a multitude of actors.

> As soon as the word 'construction' succeeds in gaining some of the metaphoric weight of building, builders, workers, architects, masons, cranes and concrete poured (or more sustainably, earth rammed) into forms held by scaffolding, it will be clear that it is not the solidity of the resulting *construct* that is in question, but rather the many heterogeneous ingredients, the long process, the many trades, the subtle co-ordination necessary to achieve such a result.
>
> (Latour [255], p.46)

'Architecture is not the work of architects' according to Fallan [220]. He asserts that 'architects are important as actors in the network that is the production of architecture but are only one group among many. Equally, engineers, contractors, consultants, masons, carpenters, electricians, politicians, planners, owners and users are key, and these are just the human actors' [220], p. 90.

Essentially then, ANT suggests architecture is the co-production of humans and non-humans. Non-technological aspects of a building are considered in a broader sense. It's not just about erecting a building, but about co-producing architecture as both nature and culture, both matter and meaning, both artefact and belief. This philosophical approach is closely aligned with the production of sustainable buildings that require a greater consideration of social, environmental, economic and technical solutions which aids the analysis of the case study buildings presented in the proceeding chapters.

According to Lee [128] there is much ambivalence towards issues of sustainability within architecture. Some architects are very committed to sustainable practice, some are very sceptical about it and some, while sceptical of the oversimplification of ideas of sustainability, are taking a thoughtful approach to exploring the concept and its implications in more depth.

Often the architect is spoken of in singular form whereas an architectural office can be a large organisation comprising of many individuals. Architectural prizes are awarded to the head of the architectural office even though the project will be an amalgamation of many people. This has resulted in the evolution of the 'starchitect' that we now associate with iconic buildings as personal expressions of architectural philosophy. They are considered celebrities and venerated as the ultimate creators of structures that inevitably university graduate architects seek to emulate. Architecture is portrayed as an autonomous 'objet d'art' and the architect is eulogised as the sole creator.

The personality cult is still prevalent both in architectural history but also to a large degree within the (contemporary) architectural community. These conventions can be challenged and explain to some degree why the architect himself has been isolated and viewed as the personification of architecture taking on a heroic appearance. The transparency of the architect's thought and process is key in the adoption of sustainable building practices and the traditional role of the architect needs to be challenged and transformed if more sustainable building practices are to be adopted, this is arguably most advanced in the field of sustainable architecture because of the need for collaboration.

Colomina [256] suggests that architectural critics and historians are shifting their attention from the architect as a single figure, and the building as an object, to architecture as collaboration. Attention is paid today to all professionals involved in the project: partners, engineers, landscape architects, interior designers, employees and builders. With this shift, methodologies of research necessarily change. This is particularly true of sustainable architecture requiring a holistic understanding of traditional, new and emerging sustainable practices, materials and technologies.

> There is no class or style of design which is unequivocally sustainable architecture, and no fixed set of rules which will guarantee success if followed. Rather, there are difficult interrelated decisions to be made and that architecture should be looked at as a cultural product that needs to be judged as an integrated entity while recognising that it is simultaneously coming from multiple origins or objectives. Buildings cannot be pure expressions of sustainability because that is never the sole objective. Indeed a true expression of sustainability may often be not to build at all.
>
> (Williamson et al. [20], p. 127)

Warwick Fox uses the term 'responsive cohesion' to describe 'a state in which the various elements of a process (design, construction etc.) exhibit a reciprocal interaction between elements that constitute the process, and the context in which it is located' [257], p. 225. Fox argues that upholding the principle of responsive cohesion in sustainable architecture entails responding to ecological, social, and built contexts in that order of priority. The responsibility of the designer is to create a design that exhibits responsive cohesion in all these contexts weaving together the ethical, human, scientific and aesthetic.

Certain software technologies are catching up to these social developments within the architecture, engineering, and construction industries. Building information modelling (BIM) promises to connect formerly disparate disciplines more intimately through streamlined communications and computational abilities.

Williamson et al. conclude that 'integrated design needs to focus on processes followed rather than design results—reconceptualisation of architecture in response to a myriad of contemporary concerns about the effects of human activity' [20], p. 136.

It has been clearly established by Bell and Fisher [258], among others, that behaviour and environment mutually affect each other and Zeisel [148] states that: 'environment-behaviour researchers need to participate in design decisions as part of the design team in an attempt to put greater emphasis on building users and their effect on the performance of the building as well as the effect of the building on the (pro-environmental) attitudes and behaviours of the users' [148], p. 53. This suggests that wider involvement of the social psychology disciplines, and in particular environmental psychologists, in the provision of the built environment is needed and more integrative design practices should be adopted.

Bennetts and Bordass [7] suggest there is lack of understanding by designers for the quantitative targets they set out to meet as well as the impact of key design strategies. They cite that some architects rely heavily on other consultants, including services engineers, to come up with the figures, effectively abrogating their responsibility for producing genuinely integrative design solutions. They suggest that designers should be able to understand quickly how a design change will affect energy use and CO_2 emissions and what it is worth in capital cost terms so that good ideas are not stripped out just because there isn't a robust argument to justify them. When designers develop their designs in a way that assess the impact and benefit of what they wish to do, science and art come together.

The Bishop Review highlights a move away from 'best' practice into 'next' practice and asserts that, 'the real impact of design in the built environment is felt not only through major projects but the design of everyday places and buildings. It is from the cross-over of architecture, landscape, engineering, construction and environmental technologies that 'next practice' will emerge' (Bishop Review [259], p. 5).

Systems and Complexity

The involvement of a wide range of participants in identifying and evaluating design problems can lead to a deeper and more extensive understanding of the client's requirement, especially where complex sustainability issues are involved. It can also lead to positive outcomes where single investment of resources can deliver multiple benefits if properly applied. To work, it is necessary to apply 'systems thinking' to construction projects to know exactly where and when to intervene to gain maximum leverage.

As discussed in Chap. 2, the way humans generally tackle complex situations is to try to break them down into manageable parts. This reductionist approach has its strengths but critics also highlight the limitations of this predominantly scientific view. This section investigates the intersection between building design and sustainability and how knowledge about buildings can be structured into a logical world view based on the theory of system behaviour. Systems psychology is a branch of psychology that studies human behaviour and experience in relation to complex systems.

It is an approach in which groups and individuals are considered in *homeostasis* (a stable and relatively constant condition). What stops people changing unsustainable habits is a desire to maintain stability, we view change as undesirable unless it is very slow. Systems theory deals with relationships between parts. Williamson et al. state that 'although it is inherently reductionist, the emphasis (of systems theory) is on trying to understand these relationships rather than simply the reduction into parts' [20], p. 81.

In a similar way the view of an entire building may be disaggregated into physical and societal processes that form complex subsystems of the whole system. A building may be regarded as an open system [260] interacting with other systems or environments external to itself. Williamson et al. [20] suggest that the common aim of sustainable architecture is to create more closed systems in buildings, by feeding back, for example, through recycling and therefore minimising the import of materials and export of waste. The input and output rules are easier to satisfy because there are less of both.

The triple bottom line of economy, society and environment represents three conceptual subsystems applied to sustainable development which are useful for general development issues but when actual buildings are considered two other subsystems appear absent—the building and the building users. A suitable building system model should be able to deal with the requirements of all stakeholders both human and non-human. Again, Williamson et al. provide a cogent example:

> The adverse impact of emissions produced by a building, such as off-gassing of materials, may have an effect on the health of the occupants, the economic subsystem in terms of productivity, as well as the surrounding environmental subsystem. Similarly, the social relevance of a building is integrally bound to how it meets stakeholder needs and its economic viability.
>
> (Williamson et al. [20], p. 85)

Bennetts and Bordass [7] recognise that one reason why buildings don't work well is their unmanageable complexity and they advise building designers to keep their designs simple and well thought out and only after that to be clever. They highlight that too often designers are rewarded for add-ons such as solar panels when much more CO_2 would be saved in other ways and at much lower cost and suggest they pay more attention to the design, installation and user interfaces of the control systems, focussing on simpler solutions.

> I can foresee passive approaches being sidelined in the rush to add renewable technologies. We can improve building energy performance by 50 or 60% over standard benchmarks by re-engineering the structure/mass and fabric/cladding. There is less to be gained from re-engineering the environmental services elements if the envelope of the building has been properly designed and engineered. (Bordass [261], p. 2).

This is known as the 'fabric-first' principle and Bennetts and Bordass [7] cite inefficient glass buildings facing south that get burdened with mechanically operable cladding systems, over-sophisticated glazing, complex lighting controls and so on, which promise to make it work but always fail in some way. There can be an over-reliance on designing-in building management systems (BMSs) and computer operated services which can be surprisingly difficult to control and change with no real audit trail on commissioning or programming. Users may not have the necessary skills, so designers should keep things simple.

Participatory Design

Participatory design is an element of integrative design which actively involves the end-users in the building design process in order to help ensure the building meets their needs and is usable. It has particular currency to planners and architects, in relation to place-making and regeneration projects and potentially offers a far more democratic approach to the design process by incorporating a variety of views with a greater opportunity for successful outcomes.

Studies in small group behaviour have produced evidence for the 'participation hypothesis.' Verba states that 'significant changes in human behaviour can be brought about rapidly only if the persons who are expected to change participate in deciding what the change shall be and how it shall be made' [191], p. 271.

Sanoff [262] a community design specialist, believes that participation allows users to feel as though they are a meaningful part of the design process instead of having a building design imposed on them, as passive consumers. Through user participation, designers can learn more directly about how people feel about design issues and how they use the building. Participation instructs users in important issues about the building process and the social, environmental and economic factors in making a building project happen. User participation positively affects the end product.

According to Sanoff, from his experiences of involving users in the design process 'the most satisfaction gained is not so much the degree to which their needs have been met but… the feeling of having influenced the decisions.' He goes on to state that 'this increases people's awareness of the consequences of the decisions that are taken' [262], p. 23.

Participants directly experience the roles of architects and planners while gaining an appreciation of the consequences of their decisions. Participation workshops also have an educational purpose, since participants become acquainted with relationships between human behaviour and the environment.

Andrea Wheeler has studied educational buildings and participatory design with children as a vehicle for encouraging sustainable lifestyle and behavioural change. She states 'participatory design can extend from very basic consultations over interior decoration, to children designing and building their own schools. Participation can range from uncritical forms of consultation to those that seek to explore power relations in depth' [263], p. 4. She cites workshops with architects where pupils were involved in planning, modelling, architecture and construction. They were encouraged how to design and how to think about the space they inhabited, the amount of room they needed and how to provide light, ventilation and avoid overheating.

Wheeler argues that many participation exercises are at best uncritical of current social and educational paradigms, the ethical and theoretical motives or backgrounds underpinning modern building methods. She states that

> The implications for policy are profound. Challenging deeply entrenched, environmentally destructive presuppositions underlying behaviour is essential. Whilst the need for lifestyle change is recognised, young people still have to reconcile such ideas with potential conflicts involving consumerist norms. If sustainable development is to be honestly and effectively encouraged, young people will need to enter into a discussion of community, relation, social cohesion and all the political and philosophical complexities this entails.'
>
> (Wheeler [263], p.13).

MacNaghten states

> the practical challenges for such (sustainable) initiatives are far-reaching and would involve sizeable shifts in the culture of planning and building practice: if we are serious in understanding the conditions for a more sustainable society, we need to recognise that the more directly involved are people in the construction and preservation of their dwellings, the more likely they are to care for and cherish the planet we all inhabit. (MacNaghten [264], p. 92)

One argument why participatory design strategies have yet to come fully to the fore is that architecture is lacking a robust and proficient knowledge infrastructure to make available knowledge more transparent and to disseminate it in a way that improves design practice and, ultimately, the quality and sustainability of our built environment, as discussed previously in Chap. 3.

4.6 The Construction Phase and Sustainable Behaviour

Considerable opportunities exist for influencing the sustainability of a building project throughout the construction phase. The combined skills, abilities, attitudes and behaviours of all operatives, from site workers, sub-contractors, site managers, architects, project managers etc., in interpreting sustainable design solutions to on-site practices can have significant implications for the environmental performance and social aspects during the construction phase and of the building itself and ultimately the sustainable behaviour of its users.

Constructors have many and varied roles in the building process. They are responsible for the procurement of many of the materials and components that go into a building. They select suppliers for the majority of these and store and process them on site. They manage the construction site, or in the case of sub-contractors, their contribution to its operation. They deal with labour procurement and community impact issues and control wastage and disposal on site.

Implementation of a design or plan is rarely straightforward. Construction from a design involves complex negotiations that may have a profound effect on the degree to which the sustainable goals of the design are realised. Inaccuracies in initial cost estimates often are at the root of such difficulties.

Construction teams must alter their structural arrangements and processes to adopt a more integrated approach for handling sustainable construction issues. Collaboration that moves beyond the linear process by which buildings have been traditionally built begins with design 'charrettes,' commonly defined as a method of organising thoughts from experts and users into a structured medium that is unrestricted and conducive to the creativity and the development of the design. These should include all engaged parties—owners, developers, contractors, engineers, architects, local community, social psychologists and educationalists.

Early involvement allows design consultants and constructors to appreciate the reasoning behind decisions and to identify the impacts in their own area at an early stage. This can contribute significantly to the profitability of a project by ensuring that everyone is fully aware of their role and responsibilities. Such an involvement also allows opportunities to be identified that might not otherwise have been considered. Greater co-ordination between design disciplines and constructors can avoid many of the common causes of delays and cost over-runs experienced in construction.

Furthermore, contractual relationships must be changed to accommodate innovation. True integrated design includes a contract whereby the owner, architect and contractor agree to share all the risk and reward according to a preset agreement. In these new contracts there is an agreement against litigation and limited provisions for dispute resolution. This will foster a cohesiveness that the parties are all in it together and less adversarial.

Contractors and sub-contractors often perceive sustainable construction methods, materials and technologies as a threat to their traditional working practices, largely in terms of increased costs and time, but also through fear of the unknown. If the opportunities are framed and communicated correctly by designers and developers they can move beyond this perception. For example, a plumbing sub-contractor may only see a sustainable building feature, such as the installation of waterless urinals, as a lost opportunity for copper piping work. A wise advocate would point out to the same contractor how the incorporation of sustainable technologies adds an entire second plumbing system when integrating grey water systems in a building.

Sustainable construction requires new technologies, materials and methods, and constructors' caution is often shown through higher pricing or outright refusal to perform the work. These actors justifiably fear that failure of equipment or systems will fall on their shoulders. Engineers or architects who are familiar with the sustainable features can reduce this fear by offering some form on indemnification for certain conditions of failure. Contractors could be provided with liability limits should the technology or system (and not the installation) fail.

Procurement

Many client organisations will have an environmentally responsible or green purchasing policy for the products and materials they use and some continue the scope of such a policy to the buildings they procure. Material selection can have a significant effect on the environmental impact of a building and the following illustrate the many opportunities available:

- Use of recycled aggregate, timber or steel may be specified in some or all of the works.
- Certain materials, such as PVC, may be prohibited in favour of environmentally benign alternatives, in all or part of the project.

- Timber may be specified to be sourced from renewable or sustainable sources, and proof of origin may be required, for instance by means of the Forest Stewardship Council (FSC) scheme.
- Materials may be specified to be supplied from local sources to minimise construction transport.
- It can be specified that there be no import or export of ground materials, therefore requiring cut and fill to be precisely matched.

Research by Scrivens [265] has shown that despite 71% of organisations stating that they have a sustainable procurement policy, only 1 in 4 procurement managers would reject a potential supplier based on failure to meet sustainability criteria. Unsurprisingly, cost reduction remains the highest priority for 89% of procurement managers, with only 52% reporting a similar level of concern for environmental issues. 2% of respondents felt that waste reduction would be the primary benefit of taking environmental issues into account when selecting materials and products, while 80% believed that energy efficiency would improve. Only 46% anticipated that there would be lower costs.

Certain requirements for construction processes can be included in tender documents. Water-intensive processes may need special attention to avoid contamination of water courses or ground water or damage to local flora and fauna as well as conserving increasingly scarce water supplies. While it is not yet likely that the construction process will need to meet energy targets, it is increasingly considered best practice to record and report how much energy and water is used in the construction process. This information can contribute to a database, which will enable benchmarks and performance targets to be specified in the future.

Tender documentation can also include requirements concerning waste. 420 million tonnes of materials are used in the construction of buildings in the UK each year [53] which accounts for 30–50% by volume of all manufactured goods, excluding food production [54]. 120 million tonnes end up as waste from construction, demolition refurbishment and excavation processes and it has been estimated that 20 million tonnes of *unused* materials end up in landfill each year [47].

Good practice in waste management is now easy to find and carry out. As information about good practice increases, contractors are starting to be required to meet targets for waste removal from site, minimum proportions of waste recycled and maximum limits going to landfill. It may be appropriate to prohibit sub-contractors from leaving packaging materials on site by meeting the cost of its removal, taking it away themselves or even not being allowed to bring it on to site.

Over their entire lifecycle, materials used in construction contribute a significant amount to the environmental impact of the construction sector in terms of the extraction of finite raw materials, their processing, transportation, manufacture into building products, packaging, installation on-site and their destination after primary use.

In order to reduce the environmental impact of building materials those responsible for creating and maintaining our built environment must engage with sustainable material issues at each stage of procurement, design, construction and beyond the design life of the materials. For example, during the construction phase alone, research by Gonzalez and Navarro [55] has shown that CO_2 emissions can be reduced by as much as 30% through a careful selection of low environmental impact materials.

The issue needs to be addressed across the whole construction cycle from the brief through to construction of the project and beyond the life of the building. This has led to the development of a range of analytical techniques collectively called environmental life-cycle assessment (LCA). There are also number of guides, resources and initiatives aimed at tackling the environmental impact of construction materials, including the Building Research Establishment Green Guide to Specification [86] and the Waste and Resources Action Plan [47] '*halving waste to landfill*' initiative.

However, research by Greenspec [87] has shown that there is a quite disparate body of existing data related to the sustainability credentials and environmental impact of construction materials due in part to the complexity of the market with a wide variety of materials and alternatives available, lack of comprehensive research in this field and often unsubstantiated manufacturers' claims. Therefore, it is difficult for built environment students and professionals alike to access clear, concise and impartial data without undertaking time-consuming information-gathering and research activities.

Selecting Contractors and Suppliers

A requirement of achieving good environmental performance during the construction phase of a building project is to ensure that contractors and suppliers are selected partly according to their sustainable performance and understanding of key environmental issues. Bidding contractors should recognise the importance of good environmental practice and the tender documentation can highlight benefits and incentives for the contractor, such as reduced landfill disposal costs from recycling, increased efficiency from better site management and improved health from lack of toxic materials. Contractors should provide evidence of their commitment to sustainability and their understanding of any unique aspects of a project. This should include many of the following:

- a company environmental policy
- a sustainable purchasing policy
- the method of implementation, including an environmental management system
- awareness of environmental legislation
- training for permanent and sub-contract staff
- evidence of environmental audit trails for their products
- evidence of imaginative community/public relations on sites
- experience of Considerate Constructors Scheme
- on-site validation procedures for material and products and evidence that such issues are understood
- method statements for waste, water and other resource management
- proposals for innovative environmental benefits to be achieved
- records of awards for environmental excellence

The role of constructors in commissioning can also have a significant impact on the ongoing operational performance and efficiency of a building. The commissioning process is one of the key areas of concern in achieving a building that operates efficiently and is free of defects. On the majority of construction sites, this process is limited by time and cost. In many cases, commissioning is insufficient to ensure that the full potential of the building is met and that systems are operating as intended by the designers. It is not in the constructors' long-term interests for commissioning periods to be sacrificed because of project overruns and it is the responsibility of the constructor to ensure that this does not happen.

Commissioning periods are often cut or run in parallel with other operations, which compromises their reliability and increases the likelihood of delays. This is costly, time-consuming and disruptive to tenants, developers and investors. It also strongly influences the success of a construction contract, and so affects the potential for future contracts.

Contractors who practice good environmental and sustainable procedures can establish prominence in the marketplace and experience business rewards from achieving high levels of sustainability. Among the specific benefits and rewards that can be achieved by following environmental good practice behaviours are:

- less money lost through waste.
- less raw materials needed.
- reduced cost of storing and transporting waste.
- less waste to disposal as landfill.
- Prefabrication and off-site working is safer than traditional methods reducing health and safety problems on site.
- income from sale of some waste, for reuse and recycling.
- less time and money spent repairing environmental damage to site.
- less money wasted on dealing with environmental legislation infringements.
- better workmanship and fewer construction defects resulting in fewer recalls for repair post-construction.
- improved corporate image and reputation, increased market appeal and market share resulting from contracts awarded on basis of environmental and social responsibility.
- less likelihood of convictions for environmental offences.
- improved communication through teamwork and involvement in design process.

Sustainable Construction Skills Gap

Arguably, the skills required to deliver low and zero carbon buildings do not currently exist in the UK and this goes right through the industry from investors, project managers, designers, the workforce and even facilities managers and those who use buildings. Bennetts and Bordass [7] suggest the current problem of lack of appropriate skills in the construction sector result from the recession of the early 1990s, when many skilled people were lost from the industry. Despite the workforce increasing since then, they argue that much of it is through casual migrant labour, which has not helped pass on skills and labour. Bordass [261] suggests that what is needed are the skills and supply chains to get the basics right.

Given the interest in earning an income as quickly as possible from a building, it is hardly surprising that construction defects feature as a major concern for investors and developers. They directly influence the availability of the building for sale or let and often sour the early relationship between tenant or purchaser and developer. Many defects arise from skills shortages and poor working environments that off site fabrication can help to overcome.

Complex building services are also a frequent cause of defects. This can have a fundamental influence on the ability of the building to function as intended. Sound commissioning is vital in this case and sufficient time should be allowed. Being the last stage of construction, it is often reduced or carried out in parallel with other site operations, which can negate the benefits.

Showing even a concern for environmental and sustainability issues may be new to a contractor, supplier or sub-contractor. In this case it will be necessary for the client or management team to establish the need for awareness training. A failure to raise awareness of environmental and sustainability issues to every member of the team could generate risks for the project, for instance leading to pollution events, inappropriate handling of waste or damage to protected wildlife habitats. A lack of understanding of sustainable construction methods, materials and technologies can lead to an undermining of the sustainable credentials of the building, for example lack of air tightness, poor workmanship, incorrect application of materials.

The UK Green Building Council [266] undertook a study to determine and understand construction industry demand for sustainability and education. The study involved over 500 construction industry professionals participating in one-on-one interviews, focus groups and an online survey. The general findings showed that there were diverse views about what '*sustainability*' is and what it means. Perceptions were that it is largely an '*overused, soft term*' but is '*here to stay and there is no turning back.*' It was noted that a general cynicism and perception exists that government has adopted '*sustainability*' as a buzz word but it is high up on the industry agenda and is increasingly impacting the bottom line of businesses.

The study went on to explain that although greater requirements to fulfil sustainability criteria exist, it is often a tick box approach—one not well regulated or established like current health and safety regulations. One participant observed that no-one checks to see sustainable wood has actually been used. The '*vicious circle of blame*' in the construction and property sector and the need to end it was cited as a significant barrier and requires the engagement of additional '*influencers of behaviour.*'

The study indicated that the industry does not feel fully qualified to deal with sustainability issues in general. Participants in the study shared a common frustration and confusion regarding how to meet both individual and an organisation's teaching and education needs and that national teaching and education provision is fragmented and disparate. It was also highlighted that the value-for-money of existing teaching and education is generally not as high as expected. Also, although there are many teaching and education providers, most take a vertical approach. For example, architects train architects and engineers train engineers. This approach makes it difficult to get an integrated multi-disciplinary holistic view. These disparities and fragmentation cause many organisations to run their own teaching and education programmes and are seeking a more structured approach with clear learning outcomes.

The report concluded that the most common barriers to the industry getting quality sustainability training and education were perceived to be: awareness, capacity and communication. There is no clear leader providing awareness of what is out there and where to go for quality teaching and education. There is a severe lack of capacity in the marketplace and this impedes delivery of sustainable buildings. Critical stakeholders have varying degrees of sophistication and knowledge about sustainability. Even when understanding is high, communication channels for information sharing are often weak or broken.

The UKGBC [266] offer a training and education programme called STEP (Sustainability Training and Education Programme) which offers built environment professionals the opportunity to enhance their knowledge and skills through courses and master classes to cover every stage of the building lifecycle. The stated intention is to enhance and help shape organisations sustainability aspirations and provide industry with the right skills for the UK's transition to a green economy.

It is critical to ensure that contractors and all suppliers and sub-contractors are aware of the environmental issues relating to the site, the construction operations, the neighbours to the site, the local authority and its requirements, awareness of environmental legislation, making available on site a selection of appropriate reference guides on environmental and sustainability issues and getting the entire team to '*buy in*' to the sustainable agenda of the project. This can be achieved by providing appropriate initial and ongoing training, by displaying posters and holding regular meetings about general sustainability issues and those which relate directly to site operations and are generally known as 'toolbox talks.' Incentive schemes can be used to reward site operatives who make suggestions for environmental improvements, such as reducing waste or improving construction methods.

Various schemes, frameworks and training programmes exist to encourage greater sustainable behaviour on site. The Considerate Constructors Scheme operates voluntary site and company codes of considerate practice, to which participating construction sites and companies register. The codes commit those sites and companies registered with the scheme to be '*considerate and good neighbours,*' as well as '*respectful, environmentally conscious, responsible and accountable*' by monitoring registered sites and companies and the display of posters around the construction site, setting out the Code to which the sites or companies are committed.

Local authorities are increasingly looking for ways that construction projects can bring benefits to their area and within the planning application provision can be made to try and employ local people on the construction site. While many of the skills needed may be unrealistic to resource locally, a large construction site can provide a good opportunity for training people. Apart from reducing the need to import workers, this can also help to improve relations between a client, their construction team and the local community.

The Construction Industry Training Board Construction Skills programme is committed to supporting the development of low-carbon skills in construction and in

2011 launched a 3-year industry-wide '*Cut the Carbon*' campaign. The chief executive, Mike Farrar stated that 'the key message… is that to achieve the government's carbon agenda and its now legally binding carbon targets, a monumental sea-change is required in the construction industry.' [267], p. 1.

The government has to ensure not only that workers are getting the right skills, but also that its own regulatory agenda and spending power, as well as any other catalysts for greener construction, are embedded in the marketplace. Farrar went on to say that: 'the skills agenda is crucial and clarity is required on exactly what qualifications will be needed' [267], p. 1. He recognises that the whole training and education system needs to *'join up'* and says that if the Department for Energy and Climate Change makes clear what skills are required, CITB-Construction Skills can work with other departments and agencies, such as the Skills Funding Agency, to influence the shape of future training through FE colleges and other training providers and can also work with local authorities.

The 'Think Low Carbon Centre' based at Barnsley College has opened a £4.2 million facility to demonstrate green technologies and provide essential skills in the run up to the introduction of the government's green deal retrofit initiative. At the time of writing, it has been reported that not a single household has yet registered for the deal [268]. This illustrates the difficulty in encouraging sustainability and providing an infrastructure for delivering it.

Barnsley College's construction department will run courses on installing and maintaining low carbon and renewable energy technologies, and on how to adapt traditional techniques to make buildings energy efficient. The centre was officially opened by the architect and television presenter George Clarke who said 'not only does this building exemplify what can be achieved through the use of new and emerging materials and technologies, but is seeking to build up a skilled workforce' [269], p. 1.

The Building Research Establishment (BRE) offer a number of sustainability training and education programmes for the construction industry from short courses and continuing professional development (CPD) to fully accredited and examined courses. A new training programme recognises that the site team needs the correct skills and knowledge to achieve sustainable buildings according to sustainable design aspirations. The role of Site Sustainability Manager (SSM) is intended to provide a mechanism to ensure that a construction site can not only be managed in an environmentally efficient manner but can also provide the site management team with the confidence and knowledge to achieve the exacting standards of the design through build quality, site operation, material use, site accreditation and site communications.

The involvement of the SSM enables an additional credit for BREEAM assessed projects to be attained. The effective involvement of the SSM is also intended to transfer the BREEAM-related design commitments through the construction phase and reduce the risk of non-compliance for a range of site and building related assessment issues. Registered SSMs are expected to demonstrate by way of audit that they have achieved environmental targets on site and effectively communicated environmental messages throughout the whole site team. They are also required to keep up to date with the fast pace of low carbon materials and technology innovations and therefore relevant CPD is required.

The '*Cafe Van*' is a project run by environmentally aware builders as a not for profit organisation. Education is delivered during work breaks from a van serving refreshments. It provides three sessions leading to the award of a certificate in environmental awareness focussing on resource efficiency and waste issues linked to the particular site and the trades working there. The sessions cover:

- Why environmental issues have become so important in construction
- The nature of construction waste and Duty of Care
- Approaches to waste management and disposal
- The benefits of improved resource efficiency
- Key environmental legislation and Site Waste Management Plans
- How construction affects carbon balance and climate change
- Practical solutions—balancing the job and the environment

80% of attendees who have received the certificate say they would change the way they work on site [270].

Ultimately, the constructors' role is to interpret the design and seek sustainable solutions that meet its requirements while providing best-value solutions with minimal environmental damage and maximum enhancement of both environmental and social aspects of the project. Much can be learnt throughout the entire construction process of innovative sustainable building but there are inherent financial and technical risks in using the process as a research and development exercise when the ultimate aim is to produce a building to fixed schedules and budgets. In operation, a building has significant potential for research and study by both building professionals and students alike.

4.7 The Post-Construction Phase and Sustainable Behaviour

Building Use and Sustainable Behaviour

It has already been established in this study that many new sustainable buildings are not meeting their proposed performance targets, particularly in terms of energy use. It is thought that one of the main reasons for this is that the end users of the building do not understand the systems installed and do not know how to use the technologies. In fact, building users frequently behave in ways that actually increase energy use above that expected, therefore occupant behaviour is a key determinant of building performance.

It has been observed from a number of studies by Baird [6], Hadi and Halfhide [7] and Bennetts and Bordass [10] that building users frequently do not use buildings in accordance with their design intent and in some cases inadvertently prevent systems from working properly or even actively override them.

Cole [119] asserts that in conventional approaches to comfort occupants are assumed to be passive recipients of indoor conditions that are maintained within narrowly defined margins by automated, centralised systems. Building performance is often invisible to the end-user who in turn is given little opportunity to control or provide feedback on their experience of the indoor environment.

Humphries and Nicol observed that when occupants have more perceived control over their indoor environment, as is common in naturally ventilated buildings with openable windows, they are more likely to tolerate less than ideal conditions.

It is proposed that technical solutions are not always the best solution. It can be cheaper and more effective to engage people in the management of the building. This idea communicates environmental behaviour across the organisation and can serve to motivate and retain existing employees and even to help recruitment as more people understand the personal benefits of working in a sustainable building, as well as organisational benefits in terms of improved productivity and attainment for the organisation.

It is possible to get staff engagement wrong sending out contradictory or confused messages, e.g. having a sustainable travel policy when the CEO drives a Bentley to work every day, fostering feelings of reactive denial in the face of self-serving denial, or providing new bike racks and showers to encourage cycling to work that are so isolated they never get used illustrating poor design for sustainability.

When communicating to building users, people do not intuitively understand phrases like *'climate change'* or quite abstract concepts like '*carbon foot-printing*' which, as a behavioural change tool, can be used quite poorly. People need to believe they can help and the problem needs to be communicated on a human scale, made personal with achievable goals. People need to understand and agree to changes before they are implemented.

Occupants are not the only users of buildings capable of influencing architecture and it is important to consider those with both direct and indirect experience including visitors, students, temporary workers, neighbours, communities, vandals, politicians, citizens and professional mediators, researchers, architecture critics, among others.

Commissioning and Handover

For the client and user of a building, the most important phase is after construction has been completed. It is at this stage that there can be a break in the chain of responsibility for the environmental performance and overall sustainability of a project. As already established, unless the project undergoes a well-organised commissioning process the intentions of the design and construction team may not be fully realised and the user may be delivered a building that operates sub-optimally.

In a study by Baird, investigating over 30 sustainable buildings it was noted that 'the importance of commissioning was stressed by several of the clients and designers and there was ample evidence of this being undertaken. In some cases, this was still ongoing, particularly in those instances where the project incorporated novel, educational or 'demonstration' features' [6], p. 20.

A traditional way of achieving this is by means of an operation and maintenance manual or user manual which often falls short of achieving full engagement of operational staff who may not fully understand the technical language used and do little to involve the users in the operation of the building. Baird's study revealed that despite the assertion that a user manual had been prepared for the building occupants, he was rarely able to see a copy.

User guides can be creative in the use of illustrations, diagrams, and narrative to demonstrate how buildings work. An architect can provide design drawings accompanied by text about the location, usage, thus raising awareness suggested by the sustainable features integrated into the building and the logic, concepts and ideas behind them. This is especially so for educational establishments. Over time, teachers and students can learn from and expand the guide with their own ideas. Thus, the user's manual 'becomes an evolving, dynamic, and ever-growing curriculum attesting to the value of the learning environment as a teaching tool. It explains how the building functions as a three-dimensional textbook for interdisciplinary learning.' [191], p. 272.

The *'Soft Landings'* approach, initiated by the Building Services Research and Information Association (BSRIA), the Usable Building Trust (UBT) and architect Mark Way [271], provides a structure for project teams to stay engaged after practical completion, to assist the client during the first months of operation and beyond to fine-tune and de-bug systems, and ensure the occupiers understand how to control and best use their buildings. '*Soft landings*' helps all stakeholders to achieve better buildings by focussing the design, construction and commissioning process more sharply on obtaining good outcomes in use. This approach helps to pass on knowledge, capture learning from experiences, reduce credibility gaps between expectations and outcomes, and provides a vehicle for post-occupancy evaluation, feedback and continuous improvement. It is in effect a set of attitudes, principles, and techniques that can be incorporated within any procurement system across different building typologies.

The Soft landings Framework document identifies five main stages which can be adopted at any point during the development process:

- Inception and briefing.
- Managing expectations during design and construction.
- Preparing for handover.
- Initial aftercare immediately after handover.
- Extended aftercare, post-occupancy evaluation (POE) and feedback during the first 3 years of occupation.

Essentially this approach aims to focus any team or organisation more clearly on outcomes and join up people and activities which in the past have been too separated. Without this integration, current trends risk fragmenting yet further both the skill base, the communication between various stakeholders in the design and construction team, the connections with building occupants and management, and the follow-through from construction into operation. In turn, this inhibits the feedback that is essential for innovations to be directed purposefully at what needs improving.

This approach is hailed as the best opportunity for producing zero-carbon buildings that actually meet their design targets shifting the emphasis for good performance away from just design aspiration to the way buildings are actually managed and maintained, creating a virtuous circle offering opportunities for truly integrated and robust design.

There is a strong need to transfer an understanding of the systems to a team that can implement them effectively. The knowledge and skills for this are becoming more specialised and the appointment of a specialist sustainability manager is becoming more common place with an understanding of economic, social and environmental sustainability in the use and operation of a building.

The Case for Sustainability Specialists

At present, several roles within an organisation may exist that engage with a diverse range of sustainability issues, often resulting in an uncoordinated approach possibly resulting in a duplication of responsibilities, contradictory behaviours, or missed opportunities for tackling waste and inefficiencies. Some aspects become an 'add-on' to an existing role, arguably leading to a dilution of the sustainability focus.

With the appointment of a sustainability specialist, variously entitled Energy and Sustainability Manager, Sustainable Facilities Manager, Sustainable Building Manager, a co-ordinated strategic approach is provided, lead by a single individual who is able to draw together all aspects of sustainability, energy management, social and ethical due diligence and with the power to delegate out responsibilities. This enables an organisation to present a clear and transparent strategy more easily for the company as a whole.

Sustainable energy management is the process of maintaining, controlling and conserving energy in a building or organisation. The role works to change organisational behaviour by adding energy efficiency and conservation into the formula of considerations for new construction, operations and maintenance and advocate for the owners and operators of facilities to think and behave more sustainably. At the same time sustainability and energy managers can conduct employee education campaigns that teach employees about the impact their behaviour can have on the environment at work, in their communities and at home.

Vicky Kenrick, a sustainable recruitment consultant states 'it stands to reason that strategic managers would increasingly seek to implement sustainable strategies in the decades to come and the most effective method of achieving this is to take energy and sustainability into the boardroom' [272], p. 14. This new role can act strategically within an organisation to help create large cost reductions as well as ensuring the organisation complies with the most up to date regulations.

Well-informed sustainability managers are vital to close the feedback loop from performance in use, back into briefing and design, and then to help manage design intent through to specification, construction and commissioning process and on into use.

Monitoring Environmental Performance

The operation of buildings is now on the point of a culture change. As buildings are being designed to run on less energy, it is becoming essential that its longer term running and operation are linked more directly to the work of the design and construction team by involving the operators and users during the design and construction stages of the project and involving the design team into the post-occupancy phase of the building's life.

The collection of appropriate data and information is crucial to the effectiveness of this process. This is a vital part of the feedback loop by which the actual performance and the causes of any discrepancies can be diagnosed and rectified.

By installing sophisticated building management systems (BMSs), it is now theoretically possible to monitor the performance in some detail and without great cost, which can be applied to improving performance and enable learning from past experiences. The following strategies can be adopted to facilitate this:

- Specify separate metering of different parts of the building (zoning). This also gives extra credits in BREEAM assessment.
- Collect information from BMS on energy, occupation, ventilation, temperatures, humidity etc.
- Appoint building services engineers to attend periodic meetings with sustainability/facilities managers and maintenance staff to check if the building is performing optimally and if not, recommission the building.
- Undertake surveys of building users.
- Discuss with design engineers the extent to which building users are not behaving as presumed. Identify whether the best solution is to educate the users or recommission the building services, or both.

BREEAM In-Use is a scheme to help building managers reduce the running costs and improve the environmental performance of existing buildings. Operating a building represents a major economic, as well as environmental and social cost, with rising energy prices and an uncertain economic outlook, cutting energy waste and other such measures can improve profitability as well as sustainability and can enhance the corporate social responsibility (CSR) profile of an organisation. One of the problems with existing building stock is that gathering and evaluating environmental performance information is complicated. There are a myriad of design variables and potential alterations that it is hard to identify the most cost-effective measures to improve environmental performance.

BREEAM In-Use evaluates operational aspects of the building and performance in use and compares this with the potential performance of the building against current best practice. It is designed to use existing data on energy and other aspects of environmental performance and highlights areas of potential improvement for both management of the building and opportunities for physical upgrading. The aim is to better inform those who occupy buildings of their carbon footprint, allowing them to enhance and demonstrate improvement against credible measures. It also allows for property assets to be evaluated against future environmental legislation—providing a measure of quality assurance in the expanding arena of sustainable development.

Post-Occupancy Evaluation

Bennetts and Bordass [7] asserts that architects who focus on iconic designs often lack the objectivity required for assessing sustainability. He blames part of the problem on a lack of culture of learning from buildings in use, either by architects, clients or contractors, and the issue is given little emphasis in design schools. Bordass goes on to state that in a review of low-energy office buildings only about a third of the energy consumption had been anticipated in the design. He states that, 'too many iconic buildings that claim to be green aren't. It is only by assessing them that we will learn how to improve designs. An awful lot can be achieved through improved control, management and user behaviour' [7], p. 9.

There is growing evidence that the functioning of a building should be evaluated from the human and behavioural point of view, as well as from a purely technological standpoint. In a post-occupancy evaluation (POE), the building occupants (users) and facilities/sustainability managers (operators) are asked how the building performs, from their point of view. This provides the baseline data from which improvement programmes can be prepared for the development of an organisation's property and facilities, and the quality of service provided to users. A responsible organisation will also report this element of its social and environmental performance as part of its annual sustainability report to stakeholders.

Blyth and Gilby describe POE as 'a way of providing feedback throughout a building's lifecycle from initial concept, reviewing the process of delivering the project through to occupation, a review of the technical and functional performance of the building during occupation. The information from feedback can be used for informing future projects' [273], p. 8. They highlight several short term, medium term and longer term benefits of POE.

Short term

- Identification of and finding solutions to problems in buildings
- Response to user needs
- Improve space utilisation based on feedback from use
- Understanding of implications on buildings of change whether it is budget cuts or working context
- Informed decision making

Medium term

- Built-in capacity for building adaptation to organisational change and growth
- Finding new uses for buildings
- Accountability for building performance by designers

Longer term

- Long-term improvements in building performance
- Improvement in design quality

- Strategic review

Three stages of review have been recommended by Blyth and Gilby [273]; the *operational review* to be carried out 3–6 months after occupation, a *project review* carried out 12–18 months after occupation, and a *strategic review* carried out 3–5 years after occupation. A number of questions should be addressed by those carrying out the POE at each stage:

- Does the building perform as intended?
- Have the user's needs changed?
- What problems need to be tackled quickly?
- How effective was the process from inception to completion?
- What can be learned for future projects?

It is highly possible that after any of the reviews or as a natural consequence of building use, changes will be made to the building. Putting POE on the project agenda from the start will focus the minds of the project team on how the outcome of the project will be measured and it enables the team to structure and record relevant information throughout the project.

Blyth and Gilby [273] also highlight three levels of investigation ranging from an indicative review, giving a quick snapshot of the project to highlight major strengths and weaknesses, an investigative review, which is a more thorough investigation using more rigorous research techniques to produce more robust data, to a deeper diagnostic review involving a thorough analysis which links physical performance data to occupant responses. Generally, this includes: air-handling, energy use, heating, ventilation rates, temperature, lighting levels, CO_2 emissions and acoustic performance.

In a recent article by Fulcher [274] UK Green Building Council chief Paul King was reported to warn that in the future, architects could be held responsible for buildings that fail to perform based on sustainability criteria. King suggested that under increased regulation post-occupancy proof of a building's poor performance could expose the profession to negligence claims. Fulcher states 'If we rethink it, redesign it, we are going to have to prove we've made it better…it will require more rigour, more science. It is going to require continued up-skilling of the profession' [274], p. 1. In the article King emphasised the need for architects to embrace POE and that architects must speak to people outside the profession to achieve greater integration across the industry.

In a pedagogical context, students, architects and administrators can compile and analyse post-occupancy data as an educational exercise. During programming, students can help the architect collect data through observation, site analysis, and other techniques. This research can then be applied to follow-up studies during POE measuring and monitoring building performance through interviewing and distributing and analysing questionnaires and surveys, reviewing the literature from similar facilities, making comparisons, reporting findings and making recommendations for future facilities.

Just as user guides can become a professional development tool for teachers, so can POE. According to Taylor 'many teachers are unaware of the learning potential in their classroom environment. They have had little or no experience in space planning and designing effective, aesthetically pleasing (and sustainable) spaces for learning' [191], p. 272.

User Behaviour

> A building, after all, is a context for human behaviour, and they are all unique.
>
> (Leaman [13], p. 36)

Humans have evolved to be adaptive and to react to the ever-changing environment around them. As such, building users will find a way to make themselves comfortable by adapting their environment. This was found to be the case in studies where building users clearly demonstrated a preference for buildings where they had control of the indoor environment [149].

From the study by Baird [6] it was noted that there was a growing acceptance by occupants of a wider temperature band and tolerance for internal thermal conditions to change gradually in accordance with the seasons. He also found that occupants appreciated being able to see or feel the effect of their operating any of the control systems to which they had access, particularly related to the growing number of buildings with full fresh-air ventilation systems incorporating devices designed to enhance the flow of natural ventilation, such as ventilation chimneys.

Collecting information about the actual performance of a project can provide feedback to the design team and members of the construction team, so that their original design intentions can be verified, or otherwise. An organisation with a continuing construction programme can make valuable use of this information in updating its design briefs and performance specifications.

It is worth noting that operating costs, whilst significant, are often of secondary importance in most occupier organisations to the costs of staff and the value of the work they carry out. Yates [16] found that staff costs are up to 40 times as high as operating costs for a typical air-conditioned office and considerably more in less intensively serviced and more sustainable accommodation. High-quality and sustainable internal environments, which staff are able to influence some control over and develop a degree of ownership, have been shown to contribute significantly to staff productivity as well as to improved levels of staff recruitment and retention. As a result, such benefits are often worth much more to occupiers in financial terms than those relating to operational costs of the building itself.

When organisations address their workplace environmental performance, they often consider expensive projects, such as replacement heating and air conditioning systems or improving insulation. What is often overlooked is the company's day-to-day occupancy of the building and how efficiently they are using the space. Many offices use a third of their peak energy use overnight when there are few employees on site. A recent study [275] found that employees have on average 10 electrical items plugged in, including PCs, fans and mobile phone chargers. Significant carbon emission and cost reductions can often be made by simply asking employees to turn off or unplug items they use every day before leaving.

According to Leaman [13] the best buildings from the users' point of view are those that respond to requirements quickly. When users operate building controls like light switches, windows, blinds and so on, users like it best when it is immediately self-evident to the user what a particular control device does. Once operated, users like it when the device gives some rapid feedback. It is then clear to the user that the device has, in fact, made a difference. In situations where users understand what is supposed to happen, they will tend to be less critical if it does not happen and more tolerant of faults. Thus, the clearer and more understandable a control device is, the better. Users take the line of least resistance and go for the easiest option, not necessarily the one with the best option.

One of the effects of modern building, especially open-plan offices, is that they have introduced unnecessary management dependency. The technologies that supposedly save on management time and expense actually waste it. Often one set of systems may be fighting another; heating versus cooling; lighting versus ventilation, and so on. Systems often settle into the least inconvenient settings. Lights on/blinds down with all systems defaulting to '*ON*' is common [13].

The least efficient state environmentally becomes self-fulfilling and leads to unhappy, uncomfortable and unproductive occupants. Unmanageable complexity is also present in the new generation of sustainable buildings, arguably encouraged by environmental rating systems that appear to give more points to the presence of features rather than their actual contribution in reality.

Occupancy densities have been creeping ever higher and it is often assumed that green buildings can also support the same kind of densities as air-conditioned buildings. Carrying capacities—that is the number of people who can be comfortably accommodated—may well be lower in buildings that use more natural ventilation systems and have lower fan and pump power, for example.

During the 1990s a series of studies under the title Post-Occupancy Review of Buildings and their Engineering (PROBE) was undertaken investigating around 20 completed and occupied office buildings, to assess and compare their performance with that predicted by the design engineers. They were undertaken in a no-blame environment to ensure honest reporting and assessment. The results were generally surprising because many of the buildings were significantly under-performing. The results enabled both the current occupiers to improve their performance and design engineers to better understand how buildings actually perform in use.

In a summary of PROBE findings Leaman and Bordass [149] identified seven strategic themes:

1. **Meet Needs**. Occupants like buildings that respond to them.
2. **Manageability**. Don't procure what you can't afford to manage. Technical performance and user satisfaction will also suffer.
3. **Integrated approach**. Comfortable buildings can be energy-efficient and cost-effective. But only if they are made to do so.
4. **Clear and robust**. Get the essentials right. Put innovations on firm foundations.
5. **Realism about process**. Buildings are more like ships than cars. They need sea trials and fine tuning.

6. **Systematic interactions**. Promote virtuous circles. Otherwise buildings will go into circles of decline.
7. **Reflective practice**. Review everything, as contexts evolve. Don't lose sight of strategic objectives, or critical details. Use feedback to learn from your own experiences and from others.

Part of the PROBE study was the use of the Building User Survey (BUS) and selected data from a study of one of the case study buildings, the WISE Project. The surveys combine to provide a comprehensive archive on the needs and attitudes of building users including comfort, productivity, satisfaction and other aspects of perceived conditions. Although not expressly focussing on sustainability issues some of the data proved to be invaluable in supporting the findings from the primary research findings of this study.

Fundamental to any strategy to improve environmental performance of a building is to encourage the adoption of environmentally sound practices of building users through raising sustainability awareness, competence and interest. One of the ways this can be achieved is through sustainability training. In addition to making users aware of environmental issues, it can also provide useful feedback to management about what is actually happening and how objectives can be met most efficiently.

For decades, spreadsheets offered the main tool to present data from the building management system. But the benefits of using a spreadsheet are limited by how much data can be presented and how user-friendly the system is. End users now expect building automation systems to be as simple to understand as their smart phones and tablets. This has prompted the introduction of 3D graphics, photorealism and animations to building management systems. The graphics are increasingly common on public displays in schools, colleges and offices. The public use of displays makes systems more interactive, and helps to raise awareness of resource efficiency.

Digital media offer powerful ways to share information and stimulate collaboration. The vast majority of people who don't readily identify themselves as 'sustainable' can be engaged in the sustainable agenda through digital media which can offer ways to contribute to carbon and energy saving that are easy, fun and fit their idea of who they are.

Digital media tools or 'apps' that can be accessed through a PC or mobile device can incentivise people to adopt behaviours that have a carbon-saving outcome. One such 'app' developed by a firm called Carbon Culture, called 'Scrunch,' is designed to reduce demand for lighting in the evening. These tools take a very different approach to telling people what they should do to reduce carbon—they give their users something they want. Digital platforms also provide opportunities to help expert users like building managers to reduce costs through higher quality performance data, transparency and information sharing.

Green Leases

In the context of a sustainable building, a landlord may have an increased need to regulate the tenants' behaviour, depending on the overall design of the building and the landlord's goals with respect to energy usage, indoor air quality, waste reduction, and so on and the lease agreement should reflect this. A 'green' lease may have increased reporting requirements for a tenant. This could require the tenant to report to the landlord energy consumption data, monitor readings related to indoor air quality issues and equipment efficiency.

Such requirements in a lease must be drafted to first require the tenant to collect and maintain such data, as well as provide it to the landlord on a periodic basis or upon the landlord's request. Failure to do so may result in tenant default. The use of the green lease by a landlord may have some unintended consequences. For example, if the lease declares the landlord's intention to seek BREEAM certification for a building and the landlord is not successful in obtaining it, a tenant may seek a remedy against the landlord for such failure, including the right to pay reduced rent or possibly to terminate the lease.

When the tenant, rather than the landlord, is driving the green leasing process, a landlord can expect to be held to a higher standard for the features and performance of the building than is typical, and the lease may contain certain tools the tenant can use to ensure the landlord meets those standards. This could pose new challenges in the financing arena, as lenders and capital partners will also be required to understand the dynamics of operating a green building and the obligations being placed on a landlord/borrower in this context.

As more tenants begin to specifically seek out green buildings, the green lease will become a requirement and the provisions of the lease may work to impose requirements for sustainable behaviour on both the landlord and tenant.

4.8 Chapter Summary

If architecture becomes more sustainable, it is because its practices and buildings will have fundamentally become more integrated. Integrative design involves not only integrating the increasing complexity of building production but it also formally incorporates and directs the behaviour of complexity.

The power of authorship beyond the twentieth century myth of the singular architect to collaborative team structures reflects the multivariate contexts of a project. As architecture moves forward into the twenty-first century it is critical that it constructs pedagogical and professional structures that position students and professionals to expand their engagement with the integrated realities of new sustainable practices. This is vital for the advancement of the ecological, economical, social and formal basis of architecture.

The user side represents a valuable body of knowledge through their specific interaction with buildings and spaces. Users can quickly detect inefficiencies that most architects are unaware of.

For many organisations, their buildings represent a large impact in terms of their overall emissions of greenhouse gases, waste generation and transport burdens. They also impact significantly on staff satisfaction and behaviour and give a demonstrable measure of their respect for the people that they employ and the clients they serve.

The key barrier to consideration of more sustainable accommodation is a lack of awareness amongst occupiers and property agents to the opportunities that exist and to their abilities to influence them.

The reasons why people fail to use their buildings as they were designed to be used include a lack of understanding of the systems by both the end users and the facilities managers, poor communication of the services strategy due to inadequate information and training, and a lack of awareness by designers of the principles of human behaviour that govern successful system and control design.

Buildings are becoming increasingly complex to operate, both for the facilities manager and the end user. Designers of buildings, systems and controls are not always aware of the impact of their decisions on the occupant and fail to realise that if they do not adapt their design to meet occupant needs and behaviour then building users will adapt it themselves. If building users are uncomfortable, they will naturally take control of their environment, even if this means doing things they have been warned against or that may even be dangerous.

For those responsible for maintaining and operating buildings on a day-to-day basis, it is imperative that buildings are designed to be manageable in order to facilitate sustainable operation. Lessons learned from post occupancy evaluation should be disseminated to ensure best practice and support a culture of continuous improvement in order to create buildings that provide optimum resource efficiency and high levels of occupant satisfaction.

Ultimately, sustainability is about people and sustainable changes in behaviour. Therefore, positive results require continual investment in technical and behavioural training and improved communication.

The following three chapters focus on case study analysis in order to investigate the practical applications through the evaluation of selected exemplar sustainable buildings in relation to sustainable behavioural change and pedagogical aims and objectives, beginning with a chapter introducing the fundamental principles behind the development of each of the buildings.

Chapter 5
The Case Study Buildings: Sustainable Features, Methods and Materials

5.1 Introduction

This chapter introduces the five buildings selected for case study, detailing their rationale with images, illustrating the respective overall developments and their key specific design features in relation to sustainable methods, materials, technologies and design interventions that impact on pedagogical and behavioural change throughout the lifecycles of the buildings. These are:

- The Wales Institute for Sustainable Education, Machynlleth, Powys, Wales
- The Derbyshire Eco Centre, Wirksworth, Derbyshire
- The Core Building, Eden Project, Cornwall
- The Genesis Project, Taunton, Somerset
- Sidwell Friends School, Washington DC

Archival research of accessible case study data such as working drawings and minutes of meetings have also been analysed to provide additional verification of findings in the design, construction, operation and use of the buildings.

5.2 The Wales Institute for Sustainable Education (WISE) Building

Architects: Pat Borer and David Lea.

Location: Machynlleth, Powys, Wales.

Cost: £5.2 m.

Key sustainable features

J. L. Clarke, *Sustainable Buildings for Sustainable Behaviour: Theory and Practice*, Green Energy and Technology, https://doi.org/10.1007/978-3-031-98905-6_5

- Low-embodied energy construction materials such as earth and hemp.
- Timber glulam structure linked with steel flitch plates.
- Bio-composite, natural fibre technologies using hemp and lime.
- Energy-efficient glazing for maximum natural day lighting and passive heat gain.
- Minimal energy requirements: solar water heating integrated into a district heating system.
- Semi-transparent PV technologies used to provide both energy and shading.
- Biomass combined heat and power linked to the district heating system and grid.
- Biological, zero energy input sewage treatment systems.
- Green transport systems using sustainable fuel sources.

General Description

The WISE building at The Centre for Alternative Technology (CAT) as well as having impeccable environmental credentials in terms of sustainable technologies, energy efficiency, passive design techniques and use of natural materials the building will feedback to its occupants' data about their energy use and resultant carbon emissions plus the thermal performance of the building fabric in the hope that this experience will influence their behaviour (Fig. 5.1).

The building is being used as a resource to teach and learn about a wide range of environmental topics providing sustainable spaces for research, workshops, lectures and seminars as well as accommodation and restaurant facilities, all offering an

Fig. 5.1 Computer generated image of the WISE complex. *Source* CAT

experience of sustainable solutions in practice with the potential to influence and effect behavioural change.

WISE provides a high-quality training and conference facility demonstrating that linking environmentally sound buildings and business practices are possible and comfortable. WISE provides a location in which to explore practical solutions to vital changes to the way we live, work, travel and do business. The WISE building embodies everything that will be taught in it, offering a living example of sustainable technologies and lifestyles.

Facilities comprise of a 200-seat lecture hall, a large break out and exhibition space, a courtyard garden, a dining hall extension and bar, workshops, offices and 24 en-suite bedrooms.

Key drivers for the development of the building were to encourage sustainable business innovation and lifelong learning for sustainability. The basic architectural premise was that all principal rooms would be shallow in plan, enabling good levels of daylight and simple-user controlled ventilation, and for every room to have a vista. The site and indoor-outdoor relationships were key to the design intent, in effect the building was designed from the inside out. In the words of the Architect David Lea, 'the different qualities of sunlight, daylight and reflected light capture the passage of time and seasons.' 'Ecotect' environmental analysis software was used early on to ensure a daylight factor of at least three.

An innovative partnering contract (PPC2000) involved the contractors from the beginning. PPC2000 was developed as a direct result of the Government's Construction Task Force 'Rethinking Construction.' It brings together client, consultants, constructors and specialists under a single multi-party contract and therefore integrates design, supply and construction processes. It sets up a practical and clear basis for all the key players to work together, according to agreed timetables, from early design stage through to commissioning and handover.

The design reduced the amount of reinforced concrete, using high embodied energy cement, to a minimum for the foundations and only to areas where it was deemed necessary for load bearing requirements. In order to optimise sustainability, limecrete, an alternative to cement-based concrete, was specified elsewhere. Cork insulation was selected to stop cold bridging and for insulating roof decks to a thickness of 200 mm. The roofs were of a standing seam design using 100% recycled stainless steel with a very low embodied energy. This was also a very simple design for decking and insulation.

A mechanical air handling unit with heat reclamation for lecture theatre was designed to accommodate rapid fluctuations in occupancy heat gains. Elsewhere passive natural ventilation systems were integrated throughout the building. Within the design brief was the requirement for a high level of thermal comfort monitoring with feedback displays in each room for occupancy monitoring.

Internal finishes included off-the-shelf earth blocks with lime render, timber finishes and the use of natural paints.

The lecture theatre lies within an enclosing cylinder of rammed earth, which is free-standing and shows the strata of its construction. The material is a by-product of a nearby quarry, graded for evenness and the appropriate mix of particle size, then rammed down with a pneumatic press between lateral shuttering and allowed to dry. The earth mix for the rammed earth walls were analysed at Bath University to ensure it would cohere properly. Its 0.5 m thickness supports the roof and provides acoustic isolation for the lecture theatre. Its 7.2 m height, great length and 300-tonne mass also constitute a heat store, a thermal flywheel to even out changes in weather and a passive solar collector via the windows of the roof and the south elevation. Its major benefits are its low embodied energy and ease of re-use, having the same compressive strength as brick but a very low tensile strength, allowing it to be very easily dismantled.

The wood for the timber frame was sources locally where possible. External walls are made by coating the frame with a 500 mm-thick sprayed hemp/lime mix. This material is strong enough to be used as a breathing monolithic construction, precluding the common complications of multiple layers and membranes, yet achieves a U-value of 0.14 W/m^2K. It remains low in embodied energy and exploits a low-energy agricultural material.

There were major problems with on-site practices regarding the construction of the rammed earth wall and irreparable damage to joinery which resulted in delays, cost overruns and the eventual replacement of the main contractor.

Energy is provided through a mix of technologies including passive solar thermal gain, solar hot water, photovoltaic panels and direct heat from a new wood-chip combined heat and power (CHP) plant.

Ironically, the initial BREEAM assessment is not very high, because despite its zero carbon operational energy and low embodied energy materials, the building loses points due to its greenfield site, infrequent bus service and proximity to other key services.

People and Pedagogy

The WISE building is targeted at a wide variety of audiences. It provides bespoke courses and conferences for architects, town planners, energy consultants, waste managers and other professionals working in fields affected by sustainability. It provides sustainable and state-of-the-art facilities for businesses to receive training and guidance on the transition to sustainability from experts at the forefront of the discipline.

WISE demonstrates to householders that sustainable buildings can be 'comfortable' and 'beautiful' according to the majority of respondents. The techniques used for designing, building and decorating WISE are the result of 30 years of experimentation and research by the Centre for Alternative Technology. Policy makers will be able to experience comfortable, modern and sustainable living first hand while receiving training and guidance from experts.

Electricians, plumbers and technicians with prior experience in sustainable energy technology can train professionally on accredited courses, in areas such as solar water heating and electricity, heat pumps, wood fuels and water management, with experienced tutors with facilities that are at the cutting edge of sustainability in the UK. Higher education and research are two crucial sectors in filling the skills gap on the road to a sustainable future. WISE will act as a 'mini-campus' for academic courses run by the Graduate School of the Environment and in partnership with universities and colleges on renewable energy, building design and environmental sustainability. Their research laboratory is also available for hire.

More students at all levels are studying the environment. The new WISE building offers a tangible demonstration of state-of-the-art sustainable design and construction. As well as full postgraduate degrees including MSc Architecture: Advanced Environmental and Energy Studies, MSc Environmental Change and Practice: Buildings, MSc Renewable Energy and the Built Environment, Professional Diploma in Architecture: Advanced Environmental and Energy Studies and Doctor of Ecological Building Practices, students are able to take short academic courses taught and catered for in the WISE building. These include environmental building, ecology, renewable energy, water and sanitation, woodland skills and organic gardening. For young people WISE offers a glimpse into the future of building design and education. Young people can test out the building's special features on tours and in workshops and classes taught by experts in sustainability.

Central to CAT's display and demonstration philosophy is the showing of innovative and environmentally benign technologies that can be readily understood and applied by all. In the words of the School Education Officer at CAT, 'The building conveys a subliminal message to school groups. It says a sustainable future can be beautiful and comfortable (Figs. 5.2, 5.3, 5.4, 5.5, 5.6, 5.7, 5.8, 5.9, 5.10, 5.11, 5.12, 5.13, 5.14, 5.15, 5.16, 5.17, 5.18, 5.19, and 5.20).'

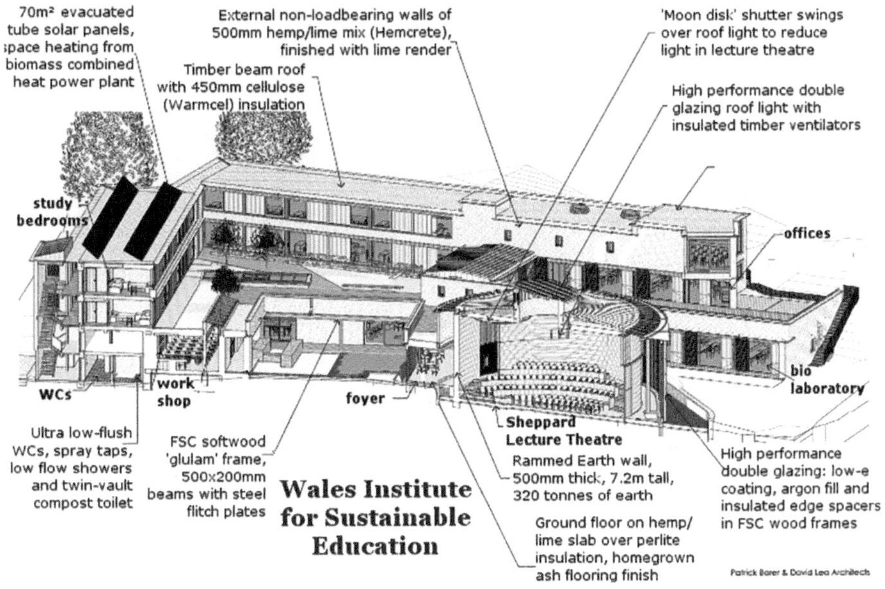

Fig. 5.2 Sustainable features of the WISE complex. *Source* CAT

Fig. 5.3 Adjusting the timber frame. *Source* CAT

Fig. 5.4 Shuttering for the rammed earth wall. *Source* CAT

Fig. 5.5 Spraying Warmcell insulation to accommodation block. *Source* CAT

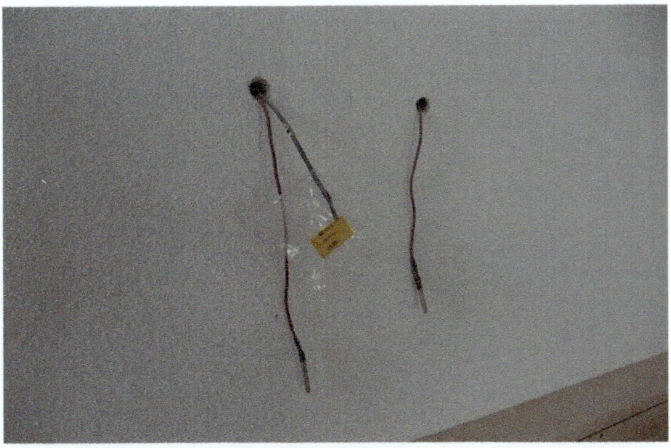

Fig. 5.6 Sensors for data loggers built in to the structure to monitor thermal movement. *Source* Author

Fig. 5.7 View from crane of auditorium and accommodation block. *Source* CAT

Fig. 5.8 Accommodation block. *Source* CAT

Fig. 5.9 Staff being shown auditorium during construction. *Source* CAT

Fig. 5.10 Students working outdoors. *Source* CAT

Fig. 5.11 Glazed corridor/buffer zone outside rammed earth auditorium. *Source* CAT

Fig. 5.12 Glazed corridor/buffer zone outside rammed earth auditorium. *Source* CAT

Fig. 5.13 Auditorium/lecture theatre. *Source* CAT

Fig. 5.14 MSc architecture: advanced environmental and energy studies lecture in progress. *Source* CAT

Fig. 5.15 Students experimenting with rammed earth techniques. *Source* CAT

Fig. 5.16 Texture of rammed earth wall. *Source* CAT

Fig. 5.17 Translucent solar array. *Source* Author

Fig. 5.18 Solar array integrated with roof structure. *Source* Author

Fig. 5.19 WISE in the context of its surroundings (winter). *Source* CAT

Fig. 5.20 WISE in the context of its surroundings (summer). *Source* CAT

5.3 The Derbyshire Adult Community Education (DACE) Eco-centre Building

Architects: Derbyshire County Architects.

Location: Wirksworth, Peak District, Derbyshire, UK.

Cost: £1.4 m.

Key sustainable features

- Solar photovoltaic cells which track the movement of the sun. Phase 1 providing 1.5 kWp with provision for phase 2 producing a further 15 kWp.

- Rainwater harvesting into a 6000 L water tank providing water to the toilet facilities.
- An air source heat pump provides energy efficient low-grade heat for the entire building via an underfloor distribution system.
- Heavily insulated envelope using recycled mineral wool.
- Some internal walls use enviro-masonry building block made in the UK from over 65% recycled aggregate.
- The remaining internal walls are constructed from a wood wool board certified as 'eco-biocompatible.'
- The external walls are non-loadbearing and are constructed from local natural Derbyshire stone.
- The glulam timber frame supports the entire roof. It is made from slow growth Siberian Larch which has a natural preservative. The glulam process adds strength and allows large spans to be covered without the use of internal pillars.
- The green roof reduces energy costs, drainage and water storage requirements and offers a wildlife habitat designed to match native local flora.

General Description

The key impetus for the development of the Eco-centre was the national drive to adopt a sustainable approach to the adult education curriculum which prompted Derbyshire Adult Community Education Service to review its facilities and curriculum. DACE decided to develop the new Eco-centre for teaching courses and as an exemplar for sustainability. The Eco-centre would help the service deliver its Sustainable Action Plan and would provide a strategy for integrating the work of the College of the Peak—a project set up to promote heritage building skills and rural crafts (Figs. 5.21 and 5.22).

Early on the client made clear that the Eco-centre should be 'a busy, welcoming and inspiring learning centre' and 'a hub for sustainability learning for the people

Fig. 5.21 Artists impression of the eco-centre. *Source* DACE

Fig. 5.22 Front elevation of the eco-centre. *Source* DACE

of Derbyshire and beyond' as well as 'a way to contribute to Derbyshire's carbon emission reduction targets.'

DACE chose the site at the National Stone Centre because of its existing infrastructure, links to other education providers, links to local employers, its central location within the county.

The Eco-centre was built over a disused lead mine which gave the team some problems which needed extensive site investigations and remedial measures before construction could commence.

Access to the centre was another issue, with limited transport links potentially discouraging groups from using it. Solutions included co-ordinating activity times to fit local bus times, encouraging sustainable modes of transport to the centre by developing the existing network of footpaths and cycle trails and promoting the use of trains as a railway station was close by. The centre was also investigating a pilot of electrically powered minibuses to transport groups from other centres.

Facilities include a large, multi-purpose area for practical work in heritage building skills, renewable energy and environmental arts and crafts. It is an adaptable teaching space with network connections for studying, meetings and lectures. It includes an outside area with canopy for practical work, an office base, a community garden, a social and exhibition area plus toilet, washroom and shower facilities.

The key design intent was that the Eco-centre is an exemplar of sustainability, both in its design and construction, and in the courses it offers. Through its design and operation, it is intended to lead by example on a range of issues relating to climate change and sustainable development. The centre and surrounding landscape is designed and built to gain a BREEAM 'excellent' rating, the first for the County Council.

Derbyshire County Council's architect team already had experience of designing sustainable buildings, so they were selected to design the centre after consultation with stakeholders, staff and learner representatives. The design team found that building an exemplar sustainable centre was more expensive than a traditional structure. This presented a particular problem in meeting the funder's specification price per square metre. Solutions were found in sourcing as much local material as possible cutting transportation costs and the carbon footprint. Costs could also be kept low by forging partnerships with local companies and offering to showcase their products in the Eco-centre in return for supplying materials at cost or in some cases for free.

The building is constructed using sustainable methods, materials and technologies including local stone, a living green roof, rainwater harvesting, air source heat pump and solar panels.

The timber frame is constructed from European Larch from a slow growth source which meant it did not need any chemical preservation and had good strength to weight characteristics. The timber beams and frame support the roof enabling a non-load bearing in-fill walling system which allows for future adaptation of the internal spatial configuration and the use of different materials, if required.

The overall design integrates well with the environment and natural assets of the site. The space was designed to eliminate internal circulation space. It has high thermal mass with high levels of insulation and air tightness.

A second phase was planned from the outset and the building is 'future-proofed' to provide further sustainable technologies with the aim of making the site self-sufficient in energy. This is dependent on securing future funding.

People and Pedagogy

The curriculum to be offered by the centre is highly innovative and new to the service, so new expertise was needed in its development and delivery. The building itself is important in promoting sustainability to learners, community groups, organisations and the general public.

In order to achieve the Sustainable Action Plan the need for key skills were identified that could be delivered at the centre. These were 'skills of empathy'—personal, community and environment, 'intellectual skills'—understanding sustainability principles, issues and finding solutions and 'technical skills.' The target audience for developing these skills were identified as families and individuals, decision-makers and leaders, employers, employees, public sector, young people, makers and installers, vulnerable and excluded people, consumers and innovators and trend setters.

The building is intended to be a catalyst for the development of an innovative curriculum. It provides courses focusing on Education for Sustainable Development, including heritage and sustainable building skills, rural crafts, arts and the environment, sustainable technologies and healthy living.

As a very visible showcase of best practice in sustainable design, construction and operation and a learning resource in its own right it is a stated intention that the Eco-centre 'motivates users to put their new skills and learning into practice' (Figs. 5.23, 5.24, 5.25, 5.26, 5.27, 5.28, 5.29, 5.30, 5.31, 5.32, 5.33, 5.34, 5.35, 5.36, and 5.37).

5.4 The Core Building

Architects: Grimshaw; Jolyon Brewis & Jerry Tate.

Location: Eden Project, Cornwall, UK.

Cost: £14.4 m.

Key sustainable features

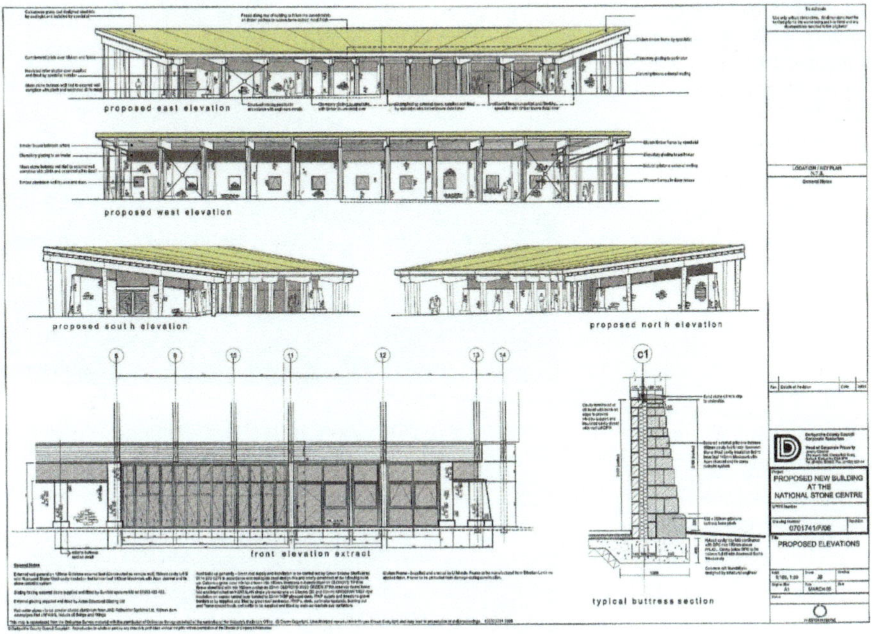

Fig. 5.23 Proposed elevations. *Source* DACE

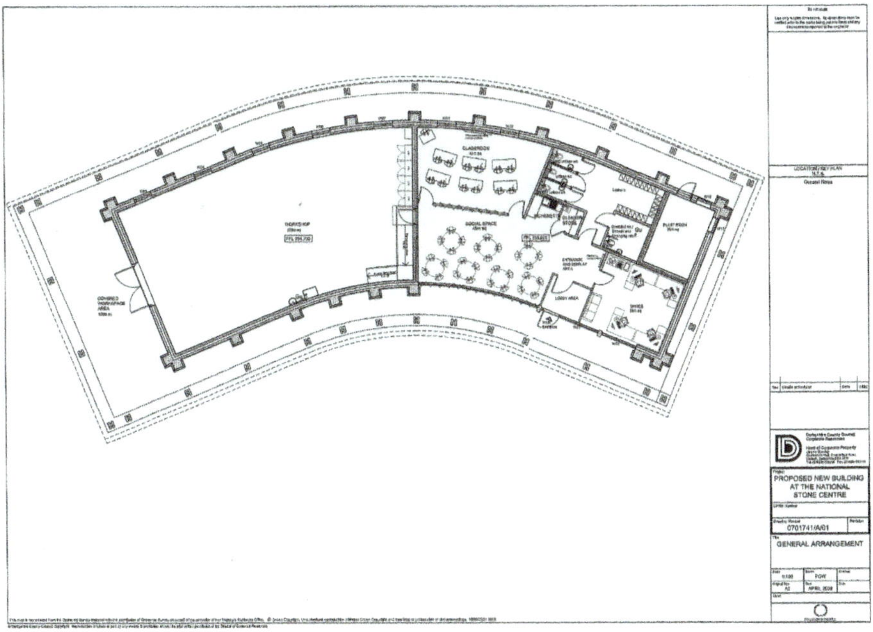

Fig. 5.24 Building plan and general spatial arrangement. *Source* DACE

Fig. 5.25 Side view of building showing glulam frame and curved design. *Source* Author

Fig. 5.26 Rear elevation. *Source* Author

Fig. 5.27 'Truth window' showing underfloor heating pipes. *Source* Author

Fig. 5.28 Underfloor heating system during installation. *Source* DACE

Fig. 5.29 Locally produced enviro-masonry blocks. *Source* Author

Fig. 5.30 As Fig. 5.29

Fig. 5.31 Solar workshop in progress. *Source* Author

- Biomimetic design (literally meaning imitating nature) and sustainable construction techniques.
- Building Management System (BMS) to regulate and optimise energy use.
- High-efficiency gas-fired boilers.
- The roofing material is constructed from sustainably sourced copper. The Core project developed a new certified supply chain for this material.
- The roof structure and frame are prefabricated glued laminated (glulam) beams from an FSC certified sustainable source. All off-cuts used as fuel.
- Rainwater collection.
- Automatic sensor water taps.
- Photovoltaics.
- Energy-efficient lighting using LED technology.
- High levels of insulation using Warmcell (100% recycled newspapers).
- Warming the air via ground tubes before it enters the building.
- A lobby reduces heat loss from the building.
- Additional electricity from a renewable energy tariff.
- Use of recycled materials as flooring: bottles, truck tyres and reclaimed wood.
- Actively investigating the future use of geothermal, solar, air-source heat, biomass and wind and hydro technologies.
- Operate a green travel plan.
- Segregation and recycling of on-site waste (45%).

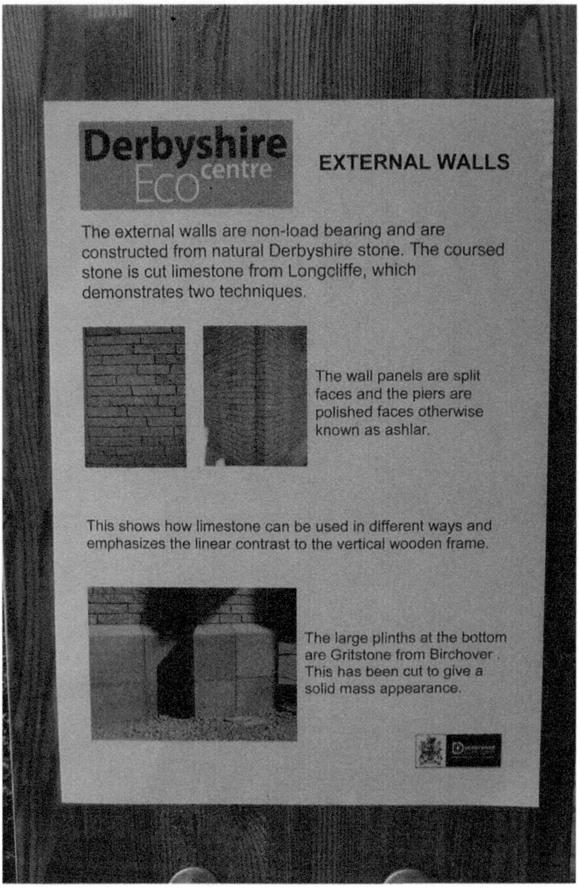

Fig. 5.32 Information board about building element. *Source* Author

General Description

The Core building was intended to be a 'test bed' for public education. It includes space for activities such as storytelling, public performance, classrooms for a full school's programme, cafeteria, workshop rooms, office space and a visitor centre housing a vast array of interactive displays including automata, machines, artefacts and models designed by 'engineered arts' (Fig. 5.38).

The Core building was designed to be an iconic building to provoke curiosity and honour the way plants power the world. The building itself is an exhibit; having taken inspiration from the tree, with a central trunk and canopy roof that harvests the sun. The form of the roof mimics nature's fundamental growth patterns in the form of opposing spirals. Based on the Fibonacci series the roof is created from an intricate web of curved timber beams, a pattern found in many natural forms including the seeds of a sunflower head, pinecones and snail shells.

Fig. 5.33 External space for practising traditional construction skills. *Source* Author

The Core is described as much a piece of artwork as it is a building and as much an expression of the botanical form as it is a space in which to tell the story of natural processes. The design for the Core was a unique collaboration between artist and architect and it is argued that it is perhaps the first time that an artist has informed the design process to such an extent on a building of this scale.

Inside the building there are three floors connected by stair and lift. The ground floor, which children can reach from outside through a tiny door and down a slide, is given over to interactive and experiential exhibitions based around the power and importance of the world of plants. The second floor branches out into spaces for exhibitions, films, talks and children's workshops. The third floor accommodates office space and a cafeteria with terrace view of the biomes and undersides of the spectacular roof structure.

A significant impact of the building beyond the site itself was the procurement of the copper for the roofing system. The US mine from which the copper was sourced has among the highest environmental and social standards of any copper mine in the world. The Core project was the first in the world to insist that all the copper used for the roofing system had a sustainable chain of custody, from sourcing to manufacture. This led to a new process of separation being used by a major organisation that is now able to offer certified and sustainably sourced copper products for the global construction industry illustrating significant organisational behavioural change as a result of working on a sustainable building.

The structure is visible and intersects through floors and other building elements allowing the building to 'explain itself.'

Fig. 5.34 Internal view. *Source* Author

People and Pedagogy

Housed within the Core are a great many innovative interactive displays encouraging experiential learning through self-directed and user-centred inquiry and play. There are also many displays of student works physically integrated in and around the building which serves to 'stitch' the building into the wider community.

The Core was intended as an education centre and hub for events, exhibitions and learning for everyone. A place to 'grow ideas, a meeting place where you can discover, learn, do, make, play, listen, talk, communicate, participate, watch, be entertained and enjoy.'

Across the Eden Project it has been recognised that staff have a significant impact on sustainability and in particular energy use in and around their buildings. Their 'Every One, Every Watt' campaign extends the energy policy to every member of staff, encouraging them to get involved in reducing as many watts of power as possible, both at work and home. The Big Green Hand awareness-raising campaign

Fig. 5.35 Visible rainwater harvesting system. *Source* Author

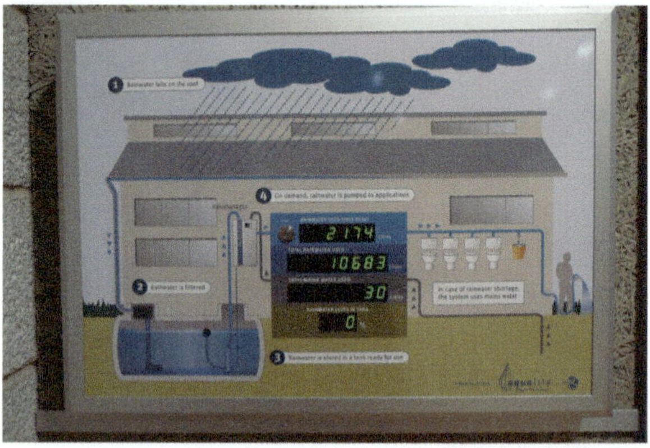

Fig. 5.36 Display of real-time rainwater collection data. *Source* Author

Fig. 5.37 Visitors view of eco-centre. *Source* Author

Fig. 5.38 The core building in the context of its surroundings. *Source* Author

gives Eden staff access to borrow a household energy meter for free and encourages them to call an energy hotline run by the Energy Saving Trust.

Eden engaged all those working on the build in the ethos of the project. Site inductions, posters, toolbox talks and training of sub-contractors as well as the Sustainability Award Scheme for sub-contractors helped involve a wide variety of stakeholders. The whole Core team took part in the Construction Skills Certification Scheme (CSCS).

Eden worked with a number of partners to share the learning experience, including Cambridge and Plymouth Universities, the Cornwall Sustainable Building Trust and Constructing Excellence. Sharing lessons learnt was a stated element of the Core's sustainability goals. Learning was shared across educational sectors: from schools to colleges and universities and Eden are working with the Construction Industry Training Board (CITB) to promote the construction industry as a career option in schools, colleges and beyond.

End-users were included in the design process and asked about their waste strategies to ensure their needs were incorporated. There was also training and induction for all workers using the Core building.

The design team worked closely with the Sensory Trust to ensure a holistic approach setting high targets for an inclusive approach to the design and management of landscape, external spaces and the building itself.

The team are proud of what they achieved in the building of the Core: from creating new social targets to the development of a certified supply chain for the copper used on the roof. They hope to 'continue to work with partners to make their buildings safe, inclusive and sustainable… and to push the boundaries in every new project' (Figs. 5.39, 5.40, 5.41, 5.42, 5.43, 5.44, 5.45, 5.46, 5.47, 5.48, 5.49, 5.50, 5.51, 5.52, and 5.53).

5.5 The Genesis Project Building

Architects: Architype Ltd.

Location: Somerset College of Art and Technology, Taunton, Somerset, UK.

Cost: £2.5 m.

Key sustainable features

- Cutaway sections that reveal the construction methods and use of sustainable materials.
- Many of the finishes are from 100% natural materials ranging from wheat fibre board to earth renders.
- Landscaping is designed to replicate natural drainage through the creation of wetlands and ponds. Pollutants are filtered at, or close to, source.

Fig. 5.39 Sustainably sourced copper roof. *Source* Author

Fig. 5.40 Key materials—stone, timber and copper. *Source* Author

- A 37 kW wood pellet biomass boiler provides the hot water for the underfloor heating.
- Each pavilion was designed to reduce the need for artificial lighting.
- A 5.1 kW photovoltaic array provides the Centre's electricity. Any surplus energy generated is directed to other College buildings.
- Retaining walls were erected with gabions. These are modular cages filled with stones which can conform to ground movement, dissipate energy from flowing water and drain freely. Silt and vegetation fill the interstitial voids and reinforce the structure.
- Furniture is made from waste wood recovered from the previous building demolished to make way for Genesis.

Fig. 5.41 Visible and tactile structural element. *Source* Author

Fig. 5.42 Visitor cafeteria under visible roof system. *Source* Author

Fig. 5.43 Over-complex user-centred lighting controls. *Source* Author

Fig. 5.44 Interactive educational displays housed within the core building. *Source* Eden

Fig. 5.45 Interaction. *Source* Eden

Fig. 5.46 The SEED sculpture at the heart of the core building reflecting the opposing spiral design found throughout nature which inspired the design of the core building. *Source* Eden

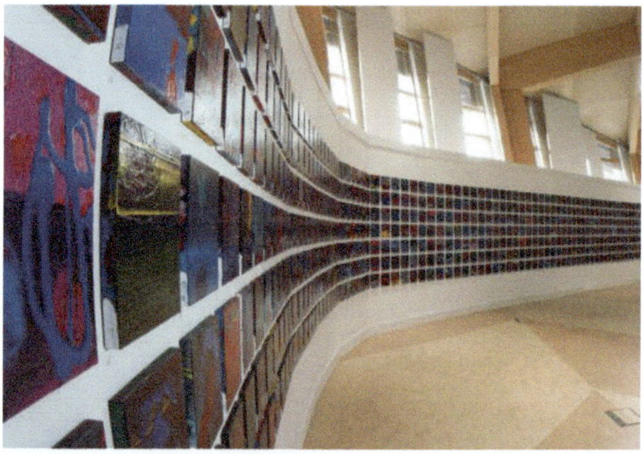

Fig. 5.47 Local artists are encouraged to display their works. *Source* Eden

Fig. 5.48 Mud and straw wall tiles made at Glastonbury festival. *Source* Eden

Fig. 5.49 Results from sustainable school workshop displayed in the core building. *Source* Author

Fig. 5.50 Mosaics by local students illustrating the water, nitrogen and carbon cycles integrated into the finishes of the building. *Source* Author

Fig. 5.51 Sensory garden around the core building. *Source* Eden

Fig. 5.52 The Fibonacci-inspired roof design with solar photovoltaics mimicking photosynthesis. *Source* Author

Fig. 5.53 The core building in the context of the Eden project. *Source* Eden

Fig. 5.54 Entrance to the genesis centre. *Source* Genesis

General Description

The concept behind Genesis was to demonstrate the many different techniques and materials for sustainable construction. The overall design had to be flexible and have the ability to stay leading-edge where necessary replacing old technology with new. It demonstrates that traditional construction materials and methods can work in harmony with innovative modern materials and techniques to create resource efficient, climate-proofed contemporary buildings. It further demonstrates the integration of 'natural' materials into mainstream new-build projects and the integration of energy efficiency measures into traditional buildings (Fig. 5.54).

In the 1970s the College had previously built, with the local education authority, a building to help the construction industry learn how to adapt from imperial measurements to metric. A great many construction workers were retrained, utilising the building as a teaching and learning resource. After it had served its purpose, the building was demolished in 2000, and the land was available for the next generation of building. Students and staff felt that sustainability was the priority, and an area of expertise that was underdeveloped in the College.

The Genesis Project started with a student assignment carried out by the College's HNC Construction students. The students produced designs and the College, with the help of the Construction Industry Training Board (CITB) approached the Regional Development Agency with a view to turning the student project into a reality. Everyone involved could see the potential of such a resource and 2 years later building commenced. Throughout the design and construction phase the ECOS Trust, formerly Somerset Trust for Sustainable Development, helped formulate the design brief and challenge the design team.

The Genesis project has 4 pavilions built around a central core that illustrate different sustainable building methods including straw bale construction, thin mortar joint clay block construction, rammed earth construction and timber frame construction. The entire Centre is designed to readily accommodate future changes through the removal and replacement of any or all of the pavilions. The externally protruding portion of each pavilion is fully finished and weather tight, while the internally protruding sections have their fabric stripped away to reveal and demonstrate the construction.

Main Glass Pavilion

A space designed to promote the exchange of ideas, exhibiting best practice technologies, with outside space for practical construction courses. Lightweight high strength steel and glass construction facilitates high levels of natural daylight. High-tech conduit for the flow of energy, water and ventilation, that acts as an environmental buffer zone and exhibition space.

Earth Pavilion

Earth's ability to absorb and release both heat and moisture makes the pavilion well-suited to its use as the Centre's shop and information centre. This pavilion was constructed using locally sourced subsoil. Designed to demonstrate rammed earth, traditional cob construction and earth blocks, the pavilion highlights their contrasting styles and properties.

Straw Pavilion

A hybrid construction of pre-compressed bales and a timber frame, the pavilion serves as seminar rooms and demonstrates a use for an industrial by-product. Energy efficient, cost-effective construction with good acoustic and thermal insulation properties.

Clay Pavilion

Highlighting construction best practice from Europe with a view to adoption in the UK. The walls are in German fired clay blocks—a material which combines structure, external envelope, moisture protection and insulation in one, offering a rapid single skin construction. Thin horizontal (bed) joints, without vertical joints, reduces mortar requirement by up to 40%.

Timber Pavilion

Lightweight super-insulated timber frame construction is used as office space. Its construction is off-site pre-fabricated modules. Wherever practically possible, the timber was sourced from sustainably managed forests. Insulated using recycled newspaper, with rapid heating response, this construction is ideal for sporadic demand environments such as offices which require rapid heating and cooling. Also known as 'responsive' buildings.

Water Pavilion

The Centre's toilet facilities provide a working demonstration of the latest thinking in water conservation through efficiency. The pavilion seeks to promote the value in adopting the use of short pipe runs, small bore pipe, low water use toilets, waterless urinals, controlled flow taps and solar heated water.

Living Roofs

Two of the four pavilions are finished with living roofs. Extensive green roofs are visually attractive, low maintenance, improve rainwater management, thermal performance and air quality, reduce sound transmissions and provide habitat for wildlife. Brown roofs in addition are a way of replacing the unique habitat that is lost when brownfield sites are redeveloped.

People and Pedagogy

The Division of Construction at Somerset College is a Centre of Vocational Excellence (COVE), and the first in the country to have received the award with 'sustainability in its title.'

The Genesis Centre is used for the development and delivery of education and training programmes offered in sustainable construction. The Genesis Project is self-funding and offers bespoke educational programmes. The Centre serves as a facility which helps people across the Southwest of England to access information on sustainable construction and which gives the opportunity to view and experience high-quality buildings which have been constructed using sustainable methods.

It also provides learning resources and materials, in a range of media, to other training and education partners in the region.

Key audiences are primary and secondary schools, further and higher education students and Colleges and professionals from the construction industry as well as individuals involved in restoration and self-build.

The Centre has created activities for both primary and secondary school pupils that map against the national curriculum and that introduce sustainability issues, fitting with the National Framework, DFES Eight Doorways and supporting Eco School status.

It delivers activities, curriculum packages and events to promote sustainable practice beyond the initial construction remit. It explores, explains and evaluates cutting-edge thinking in sustainable construction by introducing the use of sustainable practices and materials into the mainstream construction industry. The courses offered range from 1 day introductory workshop in Greening your Business through 10 week flexible learning packages in Sustainable Construction to a 2 year full-time Foundation Degrees (FdSc) in Sustainable Construction (with Modern and Traditional Pathways).

A teacher who regularly used the building stated that 'the use of natural and healthy materials had a positive effect on the behaviour and project learning of students compared to more conventional teaching and learning spaces.'

The Genesis project has strong links with the construction industry. Professionals visit Genesis for new and innovative ideas, to find out how sustainable construction works in practice, to network and to take sustainable construction courses as well as using the space for their own projects and meetings, corporate events, product launches and exhibitions. Many construction companies and organisations have pledged their skills as partners to Genesis to provide expertise and products and sponsorship for events.

Accredited continuing professional development (CPD) for construction professionals and people involved in self-build or interested in sustainability in general are run from the Centre with speakers at the cutting edge of sustainable construction. A set of nine CPD events seeks to pass on what has been learnt about the materials and technologies used, the skills needed to work with them, the knowledge gaps that exist, relevant legislation and the research and development that is currently taking place around them. Each event explores what place the methods of construction and materials used on the Genesis Centre have in mainstream construction practice. The Centre is promoted as a visitor attraction throughout the Southwest (Figs. 5.55, 5.56, 5.57, 5.58, 5.59, 5.60, 5.61, 5.62, 5.63, 5.64, 5.65, 5.66, 5.67, 5.68, 5.69, 5.70, 5.71, 5.72, 5.73, 5.74, 5.75, and 5.76).

Fig. 5.55 External view of genesis centre. *Source* Genesis

Fig. 5.56 The timber pavilion. *Source* Genesis

Fig. 5.57 The clay pavilion. *Source* Genesis

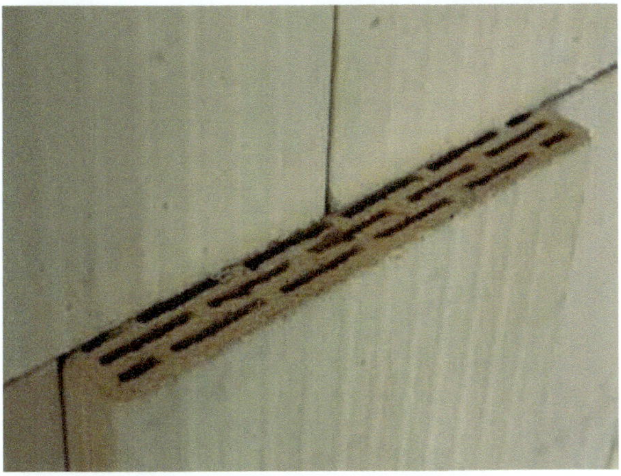

Fig. 5.58 Ziegel blocks. *Source* Genesis

Fig. 5.59 Ziegel blocks shown in-situ

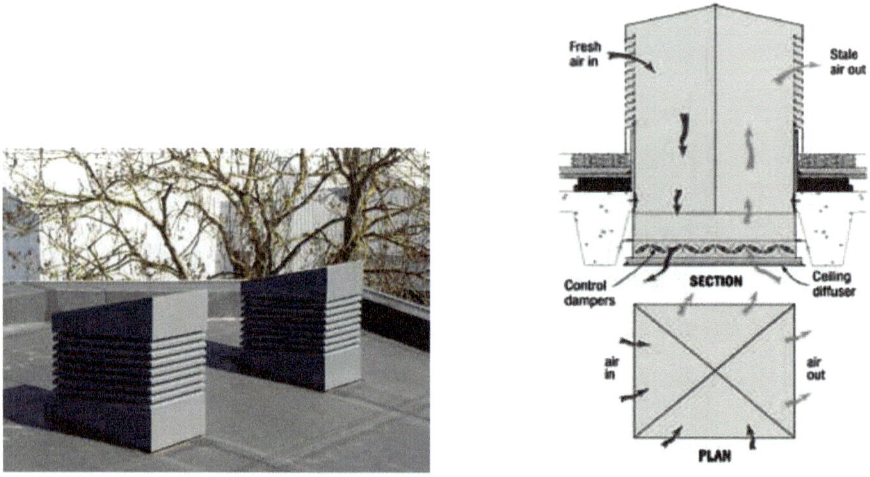

Fig. 5.60 Passive ventilation windcatchers. *Source* Genesis

Fig. 5.61 Earth pavilion showing 3 kinds of earth construction. *Source* Genesis

Fig. 5.62 From left to right: rammed earth, adobe earth bricks, cob wall. *Source* Genesis

Fig. 5.63 Earth floor finish. *Source* Genesis

Fig. 5.64 Finished cob wall. *Source* Genesis

Fig. 5.65 Finished rammed earth wall. *Source* Genesis

Fig. 5.66 The earth store. *Source* Genesis

Fig. 5.67 Earth store products for sale. *Source* Genesis

Fig. 5.68 The straw pavilion. *Source* Genesis

Fig. 5.69 The process of strawbale construction. *Source* Genesis

Fig. 5.70 The timber pavilion. *Source* Genesis

5.6 The Sidwell Friends Middle School Building

Architects: Kieran Timberlake Associates.

Location: Washington D.C. U.S.A.

Cost: Not Known.

Key sustainable features

- Constructed wetland treats building wastewater on site and recycles it for grey water use in the building saving 93% of district water supply.

Fig. 5.71 Insulating the timber pavilion with Warmcell. *Source* Genesis

Fig. 5.72 The water pavilion. *Source* Genesis

- Green roof vegetation holds and filters rainwater; gutters and downpipes direst rainwater to a biology pond, which supports native habitats.
- Building orientation, passive and mechanically assisted ventilation and solar chimneys reduce the need for supplemental energy for heating and cooling.
- Window placement, skylights and light shelves maximise use of natural light in new and existing classrooms.
- 5% of the building's total electrical load is generated by photovoltaic panels.

Fig. 5.73 Living green roof. *Source* Genesis

Fig. 5.74 Living roof run-off detail. *Source* Genesis

Fig. 5.75 Sustainable urban drainage system (SUDS) replicates natural drainage. *Source* Genesis

Fig. 5.76 Display explaining construction processes. *Source* Author

- Recycled, rapidly renewable and locally produced materials such as gypsum, linoleum, bamboo and agrofibre board are used as finishes in the building.
- Reclaimed greenheart timber is used for interior wood flooring and exterior decking; copings, site walls, and walkways are of recycled stone.
- Exterior wood cladding and sunscreens are made of reclaimed western red cedar; vertical sunscreens are oriented to balance thermal performance with daylighting.
- Paints, carpets, and adhesives are selected for low emission of volatile organic compounds.
- Energy performance is optimised with energy-efficient lighting with daylighting and occupancy controls, HVAC economiser, demand-controlled ventilation, variable speed fans and pumps, energy recovery and high-efficiency boilers and chillers resulting in 60% energy savings.
- 60% of the waste generated during construction was diverted from landfills and recycled.

Fig. 5.77 The school with constructed wetlands in the foreground. *Source* Author

General Description

The school was founded on Quaker principles, which includes a dedication to environmental stewardship. A need to expand capacity was a catalyst for the school to enhance its curriculum with an environmentally responsible focus enabled through the renovation of existing buildings and a new 39,000 square feet addition reflecting an aesthetic derived from environmental and ethical principles. The decision to construct a new Middle School, choosing sustainable design principles, was a logical expression of its values and a reinvigoration of its connection to Quaker traditions (Fig. 5.77).

The architects are firm believers in building behavioural modifications into architecture and the Sidwell Friends School project was an ideal opportunity to combine their designs for sustainability with the pedagogical needs of the school.

The design team oriented the building to take advantage of passive solar design, and light shelves and shading devices deflect daylight deep into the building while preventing glare and unwanted heat gain. High levels of thermal insulation, combined with operable skylights, windows, and cooling towers that passively exhaust hot air, eliminate the need for mechanical cooling on all but the hottest days.

If windows are opened in certain classrooms, central heating and air conditioning systems in those rooms shut down and are replaced with a system to enhance natural ventilation. Assisted by solar chimneys which heat and draw air through vertical shafts in the building, hot air rises through convection and is exhausted above the building roof.

The building features cladding made from 100-year-old wine barrels as well as flooring and decking made from salvaged Baltimore Harbour pilings.

Located near a metro stop and several bus lines, Sidwell also features bicycle storage and showers to encourage environmentally friendly transportation. Parking spaces are located underground, decreasing their contribution to stormwater run-off and the urban heat-island effect—and allowing the school to showcase more than 80 native species of plants instead of parked cars.

The new building achieved a Leadership in Energy and Environmental Design (LEED) Platinum Award, the highest rating of the four categories awarded by the U.S. Green Building Council (USGBC) and was the first platinum-rated building in Washington DC.

People and Pedagogy

The curriculum is grounded in teaching students about the natural world and their relationship to it. From the outset it was a belief that a 'green building' provides an opportunity to achieve an outstanding level of integration between the curriculum, values, and mission of the school.

Teachers at all grade levels have access to the project's landscape and building systems, and many have designed lessons around this opportunity. Students grow vegetables and herbs for the cafeteria on the green roofs, which also sequester rainwater and reduce stormwater run-off. The constructed wetland treats wastewater from the kitchen and toilets and serves as a living laboratory where students can learn about biology, ecology and chemistry.

Wind chimes in the air shafts allow students to hear air moving naturally through the vertical shafts.

The optimal use of daylight in classrooms is believed to enhance the health, happiness and ability of students to concentrate and learn. Sidwell Friends School is participating in a study by Yale University of green buildings investigating how the design affects health, emotional well-being, intellectual performance, motivation and social interaction. One student stated, 'the extra natural light in the classrooms really keeps you awake and enjoying the day.'

Mechanical controls, placed in prominent locations, are highlighted rather than hidden, just as monitors throughout the building give students real-time updates of the building's environmental systems.

As a result of the increased environmental credentials of the school the cafeteria serves more organic and locally grown food and pupils and staff have redoubled their recycling efforts.

The school's sustainable features are intended to teach and inspire current pupils and for those students to carry and transfer their knowledge and appreciation of natural systems into the future (Figs. 5.78, 5.79, 5.80, 5.81, 5.82, 5.83, 5.84, 5.85, 5.86, 5.87, and 5.88).

5.7 Chapter Summary

Figure 6.2 shows the responses from all the participants in the online survey and face-to-face interviews when asked what three words come to mind when you think about the building in the form of a word cloud where the size of the word represents the relative frequency of responses (Fig. 5.89).

The surveys and interviews have elicited a great deal of evidence about the impact of the five case study buildings on the attitudes and behaviours of key stakeholders.

Fig. 5.78 Vertical and horizontal solar shading devices. *Source* Sidwell

Fig. 5.79 Solar photovoltaic array with feedback display situated in classroom. *Source* Author

Findings from the investigation suggest, based on these case studies, how working both on and in a sustainable building can effect behavioural and organisational change throughout their design construction, operation and use. Broad opinions and attitudes are also considered relating to current trends in relation to the sustainable building agenda in the wider context of the built environment industry.

A great deal of evidence was gathered from the direct experiences of those closely involved with the case study buildings through their design, construction, operation and use. This has enabled the interpretation and understanding of phenomena in their social, institutional, political, economic, technological and organisational contexts

Fig. 5.80 Living green roof. *Source* Author

allowing for a deeper-level understanding of the responses and considered reflection on what lies behind them.

Having established the general design intent and sustainable credentials of each of the five case study buildings in this chapter, the following two chapters analyse the impact of the sustainable features on key stakeholders engaged in the design, construction, operation and use of the buildings. Establishing whether the aims and objectives of the organisations, in terms of building environmental performance, educational aspirations and impacts on sustainable behaviour change, have been fulfilled, highlighting best practice, opportunities for improvement and future possibilities.

Fig. 5.81 Solar chimney. *Source* Author

Fig. 5.82 Roof garden for student learning activities and cultivating vegetables. *Source* Author

Fig. 5.83 Solar reflector in corridor redirecting daylight into classrooms. *Source* Author

Fig. 5.84 Rainwater collection tank. *Source* Author

Fig. 5.85 Scale model of Sidwell Middle School displayed in foyer. A teaching and learning resource helping students and visitors to understand the building. *Source* Author

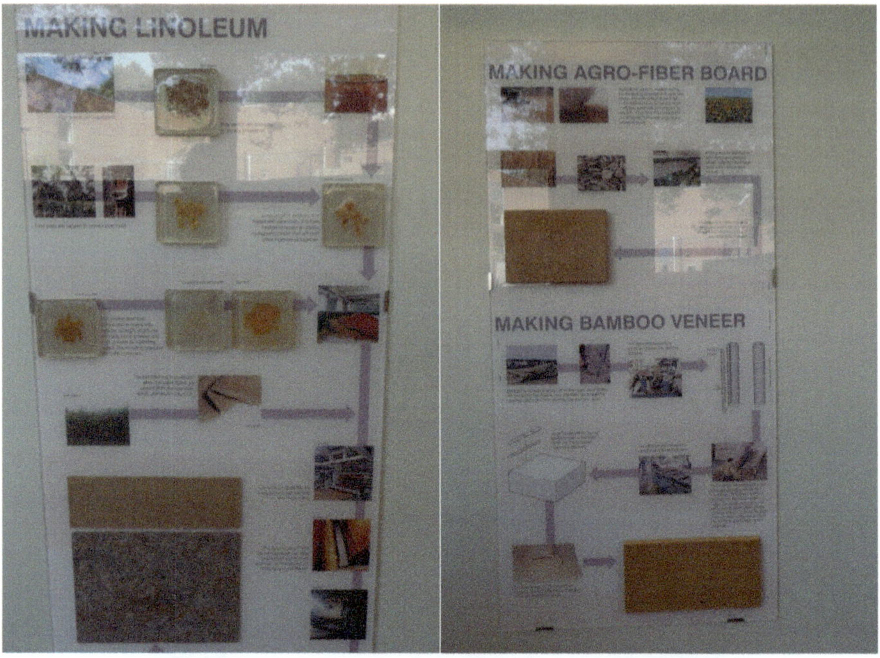

Fig. 5.86 Poster displays explaining sustainable features of the building. *Source* Author

Fig. 5.87 View of horizontal solar shading device. *Source* Author

Fig. 5.88 Water feature integrated into water management system used as teaching and learning resource. *Source* Author

Fig. 5.89 Word cloud responses to the question 'What three words come to mind when you think about the building?' *Source* wordle.com

Chapter 6
Findings from Theory and Practice

6.1 Introduction

This chapter analyses and interprets the findings from the primary research. By analysing and discussing the results based on the research findings the interrelationship of the key variables of sustainable behaviour, sustainable architecture, education for sustainable development and behavioural change theory is presented. The synthesis of these factors highlights the mechanisms for change identified from the case studies linked to findings from behavioural change psychology and are presented in terms of the overall hypothesis that sustainable buildings encourage sustainable behaviour in relation to the three pillars of economic, environmental and social sustainability.

6.2 Denial

One of the greatest barriers to behaviour change for sustainable living throughout society, and therefore present within the built environment industry, is denial that anthropogenic climate change is occurring. This enables individuals, organisations, institutions and governments to deny the significant impacts of global warming and climate change on the global economy, societies and environment, despite quite clear indications, based on empirical data, as presented annually by the Intergovernmental Panel on Climate Change (IPCC) reported in Chap. 2.

The factors that contribute to the denial that buildings and cities need to mitigate against and adapt to climate change are a mixture of psycho-emotional blocks where people perceive the threat as far away, that technological or market solutions will solve them (false positivity) and that it is the fault of others (reactive denial) and we wait for others to act before us (passive bystander effect).

J. L. Clarke, *Sustainable Buildings for Sustainable Behaviour: Theory and Practice*, Green Energy and Technology, https://doi.org/10.1007/978-3-031-98905-6_6

Other factors were found to include self-serving denial where we discount the environmental problem because of our own self-interest, often for economic reasons, citing that the problem is too big, that we lack the ability to act or our own actions are insignificant in tackling the problem or we deny our ability to perform sustainably [44]. Our response is strongest to threats that are visible, with historical precedent, immediate, with simple causality, are predictable and have direct personal impacts, whereas climate change is invisible, unprecedented, drawn out, a result of complex causes, unpredictable and has indirect personal impacts.

Traditionally, attempts to change behaviour by the environmental movement and educationalists have used rational informative and didactic techniques which have proved to have limited effects and often result in short-lived behavioural changes. Campaigns are now increasingly looking to engage peoples' psycho-emotional responses to environmental problems.

6.3 Designers and Behavioural Change

From the field studies of the five exemplar sustainable buildings the key characteristics and drivers for each of the buildings was established. They each represent a design ethos that reflects findings from the research, identified as *'design for sustainability'* (DfS), *'design with intent'* (DWI) and *'persuasive design.'* A common aim was to *'show by example'* and to influence and encourage pro-environmental and wider sustainable behaviour explicitly through the design of the buildings or implicitly through the ethos and philosophies of the organisations. The WISE building embodies what is taught within it and encourages sustainable business innovation and is designed from the 'inside-out' with the user in mind (user-centred design).

The DACE building is instrumental in delivering the local authority Sustainable Action Plan and is focussed on the development of empathic, intellectual and technical skills for sustainability and the building is a catalyst for an innovative sustainable curriculum. It incorporates features, such as the cut-away floor (Fig. 5.27) which has the design intent of raising awareness of sustainable systems, encouraging inquiry, consideration and application.

The CORE building embodies biomimicry through its nature-inspired design (Fig. 5.39) and strongly reflects biophilia, our psychological and physiological need for elemental systems and materials (Fig. 5.40). It is a testbed for public education and experiential learning through its interactive displays (Fig. 5.44) and encourages social engagement by involving local groups in designing interior spaces, *'stitching the building into the community'* (Figs. 5.47 and 5.50). The Genesis building is visible, transparent and experiential through constructional cut-aways and cross-sections (Fig. 5.59) and is focussed on mainstreaming sustainable as well as traditional construction methods, materials and technologies. The Sidwell building represents a strong moral and ethical philosophy through its Quaker heritage and the architects display a strong persuasive design agenda where a stated aim was to render the *'ethics as the aesthetic'* illustrated by the integrated solar chimney as a key sustainable feature (Fig. 5.81).

Meaningful dialogue by as many stakeholders as possible at the conceptual stage of the building represents integrative and participative best design practice. The data from the online surveys and interviews indicates that the involvement of facilities managers at the design stages of these projects shows an appreciation of the value of involving those tasked with operating the buildings at such an early and critical stage. This has implications for raising awareness, increasing knowledge and the potential for changing individual behaviour and optimising operational systems during occupancy.

Teaching staff were one of the groups most involved at design stage who are able to bring specialist pedagogical knowledge and experience applied to the methods, materials, technologies and ethos of the building itself as a teaching and learning resource utilising sustainable features for sustainable education and behavioural change. The involvement of students and volunteers at initial stages is indicated but to a much lesser extent and tends to happen more with educational buildings in the formal sector such as schools and colleges with progressive design teams and school management.

Confusion and lack of understanding about the design, purpose and function of a sustainable building with educational aspirations can have knock-on effects throughout the process of design, construction and operation. From the online survey responses, it is clear that architects and engineers are perceived to have the highest level of understanding of the purpose and function of the buildings, both achieving 100% in the range *'understood most'* to *'completely understood.'* Arguably, this is because they are closest to the design process and have the sustainable reputation of their own practices and the sustainable legacy of the building in mind, though this is not always the case.

It is interesting to note that *'visible'* sustainability was not the critical factor for at least one of the case study buildings. The main design intent was to project a more professional image which was deemed to be acceptable to corporate audiences and not to showcase its sustainable credentials. In this way the building serves a more implicit sustainable and pedagogical function which has shown to have a positive impact and suggests the possibility of buildings with high sustainable credentials moving away from the unconventional and entering the mainstream by a closer association with organisational norms of behaviour.

A deeper appreciation of the aesthetics of sustainable buildings, away from what has been identified as a trend toward aesthetic obsolescence, and a greater understanding of the embodied energy, health and environmental benefits of sustainable materials, how passive design systems work and the operation and environmental implications of (renewable) energy systems through education also impact on behavioural change. This requires a level of engagement and understanding by all stakeholders with a holistic view of buildings, structure, operation and ultimately personal responsibility for our own well-being and that of our environment.

Generally, the role of architects in projects is extremely variable and can range from highly collaborative, engaged in the project throughout to sole actor detached from other disciplines and stakeholders feeding the 'personality cult' surrounding leading architects. Resulting from this study a strong argument is made in favour of architects as actors in a network and that architecture should be a highly collaborative process defended by Fallan's phrase that "architecture is not the work of architects" [220], p. 90.

For the case study buildings each of the architects were selected for their strong environmental design and build credentials or their stated desire to engage with the sustainability agenda were seen as advocates for steering the sustainability agenda among the design teams during the pre-construction phase. The engineers selected were often specialists in sustainable technologies with expertise in using new and emerging techniques.

By involving engineers at the outset, a more integrated approach was achieved by considering how the design could accommodate the technologies as opposed to adding the technologies after the design process, commonly known as *'eco-bling'* or *'architectural greenwash'* which has been the case with many less well thought-out contemporary sustainable buildings. The project teams initially considered themselves to be working in *'a pioneering culture,'* leading knowledge and understanding, but found themselves changing the emphasis of the project remit from *'exemplar'* to *'experimental'* as the builds progressed, due primarily to practical and organisational problems encountered. This change in attitude, as unforeseen issues started to develop, speaks volumes about the fortitude of the people who worked on the case study projects.

It was stated that every decision made had to balance sustainability features against the opportunity costs of alternative and perhaps less sustainable choices. This process helped to instil 'sustainability thinking' in design and procurement decisions.

It may reasonably be concluded from the primary research findings reported in Chaps. 6 and 7, and supported by the literature review in Chap. 4, that the use of an environmental rating system can have a positive influence on the development of a sustainable building and will, by definition, have a positive impact on the sustainable practices of those involved in the design and construction process by fulfilling the requirements of the sustainable categories in order to achieve the required rating. Other building stakeholders might also benefit from knowledge of how the building fulfils the sustainability criteria of the environmental rating system, as discussed in Chap. 3.

When asked about awareness of an environmental rating system for their building, a significant number of respondents indicated *'don't know'* answers. This is a fair indication that there was a general lack of awareness of the use, or even existence, of environmental rating systems among general building users. It also reflects the difficulty in assessing and quantifying behavioural change as well as the performance of the building by general building users.

Research for this book shows that environmental rating systems do have a significant impact on the sustainability of both the behaviour of project teams and the process of development with impacts on the overall sustainability and environmental performance of the building. This could be improved if the process was made more transparent and tangible for building users.

In terms of behavioural change, those engaged with the environmental rating system were encouraged to adopt resource-efficient habits. This is a strong driver for behavioural change and research by Bhamra [136] and supports this naming the mechanism as an *'eco-steer'* or *'eco-spur.'* By adopting the environmental rating system, the project is undergoing a *'behavioural commitment'* to achieve a desired environmental rating and there is evidence that it is a powerful driving force in the consideration of wide-ranging sustainable factors often applied to working practices. Through the requirement for an environmental rating system operatives will have retained a proportion of this knowledge, according to the learning pyramid (Sect. 3.6, Fig. 3.11) of up to 75% and may well apply them voluntarily in future projects. This would be a useful question to include in a more longitudinal study, beyond the limitations of doctoral research to establish environmental rating systems as longer-term drivers for behavioural change.

The adoption of BREEAM and other environmental rating systems generally encouraged studies to be carried out to a greater level of detail than they would have otherwise been, e.g. life-cycle costing and low- and zero-carbon feasibility studies and the relocation of windows for better visual amenity. This was highlighted by one comment from a respondent: 'BREEAM was a driver to think creatively about the design of the building.'

A number of respondents indicated that some elements of the building went far beyond the requirements of the rating system and were not recognised, e.g. limited credits for use of 100% renewable energy. It was also stated that the rating system created incentives to do things that may not be cost-effective or meaningful.

The theoretical study of behaviour change mechanisms highlighted the *'mis-directed attention effect'* which when applied to the use of an environmental rating system, in the context of a number of the case studies, showed a negative behavioural outcome from an over-adherence to the requirements of the system with a number of respondents stating that it was merely a 'tick-box exercise.' This allowed important sustainability actions to be over-looked, restricting the involvement of staff members in the rating process which was identified from the WISE case study as an opportunity missed to raise awareness and understanding of the building. From the interviews it was clear that staff were generally unaware that the BREEAM rating system was being applied to the building and its implications for its sustainability.

Most deficiencies in skills and knowledge were resolved in discussion with contractors during the design and/or construction process but a number of respondents stated that some were never resolved, often resulting in poor workmanship and unplanned remedial work with additional time and cost implications as well as impacts on the energy efficiency and performance of the building fabric itself (credibility gap) with repercussions on the negative perceptions by some building users of the functionality and purpose of sustainable buildings.

Communication difficulties around sustainability issues were identified between each of the phases where the *'value-action gap'* was particularly pronounced between developer and architect, architect and contractor and architect and building users. When communication about sustainability issues were discussed the problem of the *'language of sustainability'* was highlighted and a common language for sustainability, across disciplines, was needed. This is an issue for education and training as highlighted in Sect. 3.6 on buildings and pedagogy.

In answer to the question 'To what extent could working practices have been improved at pre-construction phase?' responses generally focussed on earlier involvement of key stakeholders to *'diminish the learning curve'* and *'smooth the construction and occupancy processes.'* It was suggested that the contractors' team could have a *'design-oriented'* member to alleviate design-to-practice barriers. Also, the allocation of a budget and time allowance for pre-construction design meetings and more comprehensive briefing of the architect by the developer, enabling the formation of more detailed proposals and the use of different contractors with a more proactive approach to innovative sustainable building methods. From these responses it can be implied that communication between disciplines, more integrative working practices and challenging of set ideas (habitual norms) would have positive results on sustainability outcomes, in terms of the building and behaviours of stakeholders during the design phase.

6.4 Constructors and Behavioural Change

The transition from design to construction phase is critical for any construction project, arguably more so, when aiming to achieve high sustainability standards using unknown and often untested materials, methods and technologies. The attitudes, beliefs and behaviours of individuals and organisations in response to these challenges is key to success and highly dependent on understanding, knowledge, communication, willingness to change or adapt working practices and to undertake additional training.

Cost and time constraints were cited as major barriers to providing instruction and training, supported by findings from behavioural psychology, termed 'behavioural costs.' There was also strong evidence from case study analysis of embedded resistance to change as a result of individual and organisational norms of behaviour. From the interview data analysis, it was revealed that certain prevailing attitudes exist about sustainable buildings being problematic during construction, largely due to innovation and experimentation on-site, leading to additional costs, time and effort, already identified as *'behavioural costs,'* which tend to present significant barriers to behavioural change. This was powerfully demonstrated by the experiences of mainstream contractors working with a rammed earth system (Fig. 5.4).

Significant changes in attitude and behaviour were indicated where the respondents had previously only been involved in *'mainstream'* development and construction projects where sustainability was not at, or near, the top of the agenda. Working on the case study buildings and being faced with alternative approaches, learning curve was at its. At this point of 'unfreezing' norms of behaviour and entering the 'change' phase it is important to consolidate and reassess behaviours in order to clarify reasons for sustainable behaviours and the value of change, so that behaviours do not revert back to less sustainable norms of behaviour. This process was evident for a number of the case study buildings with evidence of both positive sustainable behaviour change, implying a move towards more long-term sustainable practices in the case of the CORE building and negative experiences where behaviour change reverted back to unsustainable practices in the case of the WISE building, citing a lack of supervision and on-site training.

The overall view of the contractors was that innovation on-site does not work well. However, research suggests that learning by trial and error, observing how others behave and modelling our behaviour on what we see around us provides more effective and more promising avenues for changing behaviours than information and awareness campaigns. Sourcing unfamiliar materials took more time and resulted in unexpected cost implications.

Comments from the construction phase interviews highlighted the need for builders to learn from one another's mistakes and successes must be emphasised so that 'the entire industry can start to gain from these experiences.' It was also noted that people have a tendency to focus on the negatives of previous projects as a reason not to continue with a new or experimental sustainable material, method or technology.

This demonstrates elements of 'self-serving denial' (the denial of our ability to perform sustainably), the *'contribution ethic'* (that we have tried, failed or done our fair share and there is no need to do more) and *'unsustainable habit reinforcement'* (confirmation our own prejudices and attitudes to environmental problems) resulting in an incomplete process of behavioural change where there is a reversion back to unsustainable norms of behaviour. This is because at the critical point of change there is a lack of positive reassessment and realisation that this is a process of change.

This requires immediate clarification by a sustainability advocate, someone who has made a conscious effort to change behavioural practices (Gestalt or Eco-psychological behaviour). This was clearly demonstrated by the failure of the rammed earth wall at the WISE building. This sentiment was corroborated by several interviewees and highlights the need to overcome such barriers.

It was stated that delegates from built environment organisations attending conferences and seminars at some of the buildings underwent awareness raising and were inspired, with the potential to change not only their own individual attitudes and behaviour but their professional and organisational behaviour applied to on-site practices, dependent on their standing and influence within their organisation.

It was cited that the use of sustainable materials, technologies and methods on site were not technically difficult but required some training and skills enhancement which would have benefitted the project in terms of quality, performance and durability of the buildings and engendered greater *'sustainable thinking'* (an *'eco-psychological state'* according to behavioural change theory) by the contractors which would have saved time, money and effort (behavioural costs) from the outset.

These results are supported by findings from literature review and in particular the field of Norm Activation Model (NAM) as discussed in Chap. 3 which essentially states that awareness and belief in the positive consequences of sustainable behaviour are least effective when behavioural costs (time, money and effort) are high. Findings from case study investigations and existing literature show that the construction industry is *'habitually strong' and* considerable barriers to behavioural change exist in the construction industry.

Applying *'habit deconstruction'* methods to the construction industry can shift these strong habits, as highlighted in Chap. 3, particularly when they target points where habits are most likely to shift, during education, training or re-organisation. This was evident throughout the case study findings, highlighted by the change in perceptions and use of certain materials, for example, the increased health and environmental benefits of switching to low VOC paints and finishes. As highlighted in the theoretical study, strategies for habit deconstruction are most effective when they are goal-oriented, successful activities are repeated, they are mediated by cognitive processes and the situation is associated with behaviour.

Construction phase initiatives that attempted to deconstruct habits and promote sustainable behaviour across the case study projects included preparatory meetings and regular site meetings with sub-contractors and a slide show for construction workers with an opportunity for questions and answer and discussion session. This supports previous research findings that linking education, sustainability and construction practices is critical at this stage in the development process for embedding sustainable behaviour. The interviews revealed that one of the most valuable construction phase activities, affecting attitudes and understanding about sustainable practices, was staff, students and visitors being shown around the site (Fig. 5.9) and in some cases experimenting with sustainable materials, methods and technologies (Fig. 5.15), true experiential learning.

Several respondents cited the *'macho culture'* of the construction industry as a significant barrier to the adoption of sustainable on-site practices where it was often perceived as socially taboo among the peer group (the cultural norm) to engage with sustainability issues. This indicates a clear need to tackle the embedded culture of some elements of the building industry.

Most interviewees in each of the projects at design and construction stages cited the need or desire for more time in order to fully achieve sustainable aims and objectives. Another recurring theme was the desire for more training. Many groups found that they could only train the onsite workers to a limited degree and that the contractor needed to take a more proactive role in training their staff. Several sites found that while the top-level officials in an organisation understood the principles behind the sustainable construction methods, many of the actual site workers and their managers did not.

The age of site workers was cited as a factor in sustainability and behavioural issues. Younger workers were less engaged with the sustainability agenda, which had not been reflected in their mainstream training or work experiences. Older workers were viewed as having a better understanding and appreciation of resource efficiency and waste issues. They did not necessarily relate this directly to sustainability, which was not fully understood as a project aim, but more to good on-site practice.

Psychological barriers identified in the literature review and presented in Chap. 3 indicate that individual contributions to collective problems are often seen as futile, widely known as the 'commons dilemma' where individual or organisational needs are pitted against the needs of the wider community. In order to act sustainably individuals and organisations perceive that economic benefits need to be sacrificed for environmental gain. Here, there are also elements of self-serving denial in the disregarding of environmental problems, discounting liability, denial of ability to perform sustainably and that small-scale actions are ineffectual in tackling large-scale problems.

In an unregulated, free-market economy, identified in Chap. 2 as *'indifferent capitalism'* this is often the case where success is based only on financial performance and responsibility is only to shareholders in a globalised, highly mobile market, attempting to achieve maximum growth, with short-term investment horizons where social, moral and environmental costs and benefits are externalised and de-coupled from economic growth and are therefore un-priced.

A shift to responsible capitalism or *'capitalism with a difference'* is highlighted in Chap. 2 where commercial liberty is more balanced with moral sympathy, social and environmental metrics are included with financial performance, business is answerable to a broader range of people such as employees and community and is characterised by self-sustaining local communities within internationalised boundaries, optimal growth, place-centric with a greater focus longer-term value. This theoretical view is strongly supported by findings from the case studies where the WISE building contracting organisation was subject to a buy-out by a venture capitalist mid-project.

This significantly shifted the business model from a family run organisation with good sustainability credentials and a long-term view of the commercial benefits of a more socially and environmentally aware approach to a model that reflected the view that protection of the environment is contradictory to economic interests. As a result, the lack of awareness and belief in the positive consequences of sustainable behaviour resulted in the perception that the behavioural costs of time, money and effort to achieve greater sustainability were too high.

Financial benefits from adopting sustainable behaviours can result in capital cost savings, reduced running costs, increased investment returns, increased productivity, staff recruitment and retention, more efficient resource use and improved marketing and corporate image. The other case studies each showed an awareness and belief in the positive consequences of sustainable behaviour more in line with the 'capitalism with a difference' model; the DACE project with its environmental and social agenda, the strong pedagogical aspirations of the Genesis project and the moral and ethical beliefs of the Sidwell project team.

In order to establish participatory working practices, the online survey asked which groups participated in the construction phase for each of the five case studies illustrating the level of involvement of certain identified key groups including community groups, students, volunteers, facilities managers, sustainability consultants and teaching staff. Involving these groups, in construction projects was highlighted as a key teaching and learning opportunity with the potential to not only change their own sustainable attitudes and behaviours but also those of the operational teams and the overall sustainability and pedagogical efficacy of the buildings during occupation. The groups highlighted as being involved the most were facilities managers and teaching staff.

The pedagogical and sustainability functions and purpose of the buildings offer significant potential for this to occur. Theoretical study revealed that the active involvement of end-users can be a meaningful and more democratic approach to the design and construction process. Macnaghten [264] asserts that the greater the involvement of end-users in the construction and preservation of the built environment the more likely they are to care for the wider environment, or in other words, increased sustainable behaviour.

Questions in the survey sought to elicit perceived behavioural and attitudinal changes towards sustainability during the construction phase based on a number of sustainability categories. *'Recycling and waste,' 'materials,' 'water use'* and *'management'* showed the highest perceived impact for driving attitudinal and behavioural change. The highest frequency of positive responses showed *'energy use'* receiving the highest response rate, followed by *'materials'* and *'indoor environmental quality,'* then *'recycling and waste,' 'water use'* and *'land use.'* Neutral responses (those that scored zero) to this question were most numerous across each of the categories with the exception of *'materials'*.

Responses that showed greatest negative impact on attitudes and/or behaviours were in the categories *'materials,' 'water use'* and *'indoor environmental quality'* with *'transportation and materials'* the lowest. It can be concluded from this that different experiences result in different perceptions and attitudes to sustainable features, particularly at construction stage where often untried and untested materials are more likely to result in suboptimal performance requiring reassessment of methods and practices.

Revealingly, 23 of the 29 respondents chose to skip the question related to the attitudinal and behavioural impacts of these buildings' sustainable features during the construction stage showing a general lack of willingness or ability to consider this question within the limits of the online survey. This was a complex question for respondents to consider within a questionnaire-based survey. It indicated under which categories of sustainability most or least change had taken place across the five case studies but did not elicit what caused the change. The question could have asked for more explicit answers to what caused the change but it was felt this type of question better lent itself to the more qualitative methodology used during the face-to-face interviews, as presented in Chap. 7.

It was found that the construction phase is highly influential in how the design solutions are interpreted to on-site practices with significant implications for performance in use and that a more integrative approach can encourage sustainable behaviours. Different contractual arrangements share risk and rewards and have proved to be less adversarial, demonstrated by the successes of the partnering contract used by the WISE project team, reported in Chap. 5.

Their procurement policy was shown to have a significant effect on the sustainability of projects in terms of the carbon intensity and scarcity of products and materials and requirements for more sustainable processes were included in their tender documents with behavioural commitments for water efficiency, the recording of energy and fuel use and waste reduction. WISE was exemplary in its collection and analysis of such data, to inform future projects, for educational purposes and wider dissemination.

6.5 Building Users and Behavioural Change

Many of the respondents to the case study research felt that there was not much personal change in their own behaviours or attitudes because they were already sufficiently informed about sustainability issues and act accordingly. This belief can be attributed to the 'contribution ethic' from behavioural change theory which in essence states that we can feel we have done our fair share and there is no need to do more, whereas further investigation revealed that in many instances the respondents own lifestyles could achieve greater sustainability by not using a car or by investing in renewable energy for their homes. Many did however admit that their knowledge of materials and willingness to use new materials grew from experiencing the project builds (Fig. 5.9).

Many interviewees noted that even with a building that is explicitly sustainable they saw a lack of change in people's behaviour within and around the building. People still did not turn off lights or recycle as much as envisaged, considering the deeply embedded sustainable credentials of their surroundings. One interviewee highlighted that a sustainable building does not, by default, encourage people to behave in a sustainable way but can provide the setting for increased sustainable behaviour, given sufficient 'eco-information' through sustainability training and education, 'eco-choice' (Fig. 5.32) with the option to select sustainable products and services, 'eco-feedback' showing users what they are doing with sustainability implications of their actions (Fig. 5.36) and with reminders to act sustainably (Fig. 5.86) and eco-technology which ensures behavioural change can take place by providing the hardware that enables change (Fig. 5.79).

A general lack of interest in how buildings are managed is a factor in how users respond to their buildings. A greater awareness of resource use and performance and the impact of user behaviour was highlighted as an area which might change perceptions and future behaviour with a positive environmental impact.

One of the key research objectives of this study is to discover the behavioural factors involved in the identified '*credibility gap,*' that is the discrepancy between aspirations of designed performance and the actual performance of a building. Occupant behaviour was found to be a key determinant of building performance which is often invisible to the end user. Users are commonly assumed to be passive recipients of buildings and indoor conditions whereas it can often be cheaper and more effective to involve and engage people in the management and operation of the building as it was found that the greatest costs throughout the lifecycle of buildings are incurred through occupancy.

This strategy can also motivate and retain employees as they become aware of the environmental, economic and social benefits. From interview data analysis a critical knowledge deficit at handover was identified, supported by findings from literature review. It was recognised that staff and other building users need time to assimilate and acclimatise to a new built environment and to experience the variability in comfort conditions and associated control mechanisms throughout each season, at least. It was stated from numerous respondents that guidance should be in non-technical easy to understand language, in written form (for reference and accessibility) and training should be rolled-out in stages to prevent information overload. This should be an iterative learning process at seasonal intervals and as stated by one respondent 'any new building takes a year to understand and perhaps a sustainable building takes even longer.'

There is a tension between technology and occupant control in achieving optimal performance of the building. We are increasingly experiencing a shift to more complex control systems (Fig. 5.43) with design responsibility moving from architects to engineers and control responsibility from occupants to relatively untested 'intelligent' technologies that promise intelligent or smart buildings, removing the need for user control. This was convincingly demonstrated for one of the case studies where movement sensors for lighting were incorrectly located so that users of toilet cubicles were not detected by the sensors resulting in them 'sh****ng in the dark.'

Leaman and Bordass [149] suggest that we are seduced by the promise of technological solutions that often operate at sub-optimal levels, not least because human behaviour is counter-intuitive, irrational and unpredictable when faced with purely technical solutions and poor design is mistakenly based on predictions of behaviour. Examples of observed irrational or contradictory behaviour from the case studies where members of staff were fully aware of energy efficiency issues included having windows open while the heating was on.

Literature review revealed that when building users are removed from control responsibility, by relinquishing control to managers, or are faced with unmanageable complexity of control systems, all systems tend to default to 'ON' leading to the least efficient state where systems fight each other and occupants are unhappy, uncomfortable and/or unproductive. An example of this was observed in one of the case study buildings when, during the day lights were on while blinds were pulled down.

Other observed behaviours included residential course participants circumventing the passive ventilation system by opening windows in the shower cubicle to alleviate the build up of steam, which resulted in the system operating sub-optimally which in turn resulted in the windows being nailed shut by the building operators during the winter. This is a good illustration of users wanting to control their own comfort conditions but being thwarted by well-intentioned but over complex systems. If the system had been explained to residents, this may have resulted in its optimal performance with more sustainable behavioural outcomes.

It is clear from observed behaviours through the case studies and secondary research findings that it is the role of building designers to support intelligent patterns of behaviour of the occupants themselves with appropriate and clearly understood systems. From the surveys it is clear that users appreciate being able to see and feel the effect of their operation of systems. Users instinctively adapt to their environment and prefer to control it themselves and it was found that from a users' point of view the best buildings in terms of operation and performance are those that respond quickly to their adjustments and offer tangible feedback. Under-floor heating was used in a number of the case study buildings (Fig. 5.28) as it offers energy efficiency and greater thermal comfort but its operation was generally not understood by the building users, being slow to respond to adjustments and users reverted to familiar behaviours for traditional gas-fired heating systems by turning the system up to full capacity when it did not respond immediately. Here, education, training and experiential learning is key, as discussed in Sect. 6.8 below.

Across each of the five case studies the surveys showed convincing evidence that the design aims and objectives were achieved in terms of the *'triple bottom line'* of sustainability, namely social, economic and environmental criteria with a high proportion of respondents believing that the environmental objectives were achieved. Not unsurprising, as this was a key driver for the development of each of the case study buildings.

Greater *'eco-feedback'* for both the behaviour of occupants and the performance of buildings has been shown to have positive effects and serves to link building users to performance criteria, particularly in relation to financial savings. Although building users and visitors tend not to be responsible for running costs it would increase their level of awareness and arguably impact on their behaviour as responsible building users. Post-occupancy training was raised as a significant issue, due to the general reliance on oral training of individuals, a lot of the building management information was lost as it moved on with changes in staff and the building's performance-in-use suffered accordingly.

One of the main discoveries, from overall research findings, is that the level of personal user control over heating, cooling, lighting and ventilation systems has a significant impact on actual and perceived health and well-being, comfort levels and individual empowerment as an influential actor in the operation of the building. The findings from the primary and secondary research show this to be a critical factor in the process and performance of the day-to-day running of the building.

The opposing view is that *'intelligent buildings'* remove individual control in favour of automatic regulation of building systems. It was found that lack of user control can lead to disempowerment and a heightened sense of dissatisfaction within a narrower range of comfort conditions. Another issue that arose was the significant and consistent failure of building management systems (BMS) to operate at optimal performance either through technical issues, such as poor installation or system failure or operational issues because of over-complexity (Fig. 5.43) or lack of training in their usage.

As highlighted from literature review [12] and from the primary research findings, there is often a significant gap between predicted performance at design stage and actual performance-in-use (credibility gap). The perceptions about the performance of the building by users and operators are important in relation to the original design intent in terms of how individuals interact with the building and its ongoing operational systems. If users perceive the building as wasteful in terms of resource use, they are less likely to attempt to behave sustainably, regarding their own individual actions as futile. Even when there is an intention to act prudently, such as to invest in energy-efficient technology, the resultant fall in energy bills, because of improved building performance, can lead to increased consumption as an unintended consequence, identified from literature review as the *'rebound effect.'*

However, there are numerous exceptions across a wide variety of building typologies and variances in sustainable behaviour where a building can perform well in sub-optimal situations. For example, some poorly insulated buildings can have a low environmental impact through prudent user behaviour, such as simply putting on extra layers of clothing or a highly insulated building may offset poor user-behaviour by eliminating the need for heating.

6.6 Strategies for Behavioural Change

The respondents' perceptions of the drivers for the production of the case study building's, as indicated by responses to the post-construction element of the interview survey, strongly reflects the triple bottom line of sustainability, showing strong awareness that the projects were driven relatively evenly by economic, social and environmental criteria. These were established by the 2005 World Summit of the United Nations (and identified throughout this thesis) as critical factors for approaching a balanced progression towards sustainable development and behaviour change reflected in the design, construction and operation of these buildings.

In some cases consequent strategies, encouraging change after the targeted behaviour, were incorporated such as eco-feedback of carbon savings as a result of the intervention of eco-technologies (Fig. 5.79). For building occupiers focussed behavioural expectations and goals (behavioural commitment) were encouraged, with rewards if energy savings were achieved for each department. The Eden project's *'Every One, Every Watt'* campaign extends the energy policy to every member of staff, encouraging them to get involved in reducing as many watts of power as possible at work and at home, illustrating how the organisation can engage in sustainable behaviour with individuals fostering a culture of collective behaviour, tackling the identified perceptions that individual actions are futile in the face of environmental problems. Working in peer groups and sharing learning is a proven strategy for bringing about greater behavioural change.

Across the five case studies it has been shown that staff can have a significant impact on sustainability if they are engaged and informed about the building's operation and in particular energy use in and around their buildings, with the opportunity to bridge the credibility gap between design aspirations and performance in-use.

It was shown that site inductions, *'toolbox talks'* and training of sub-contractors, representing a combination of behaviour change strategies; *'eco-information,' 'eco-choice,' 'eco-spur'* and *'eco-education,'* encourage the sharing of the learning experience. This is strongly supported by behaviour change work by Global Action Plan, a leading behavioural change charity, which advocates facilitated peer group working (Fig. 5.14).

This encourages participants to challenge existing strongly held habits in a supportive environment and make new connections between their own familiar working or lifestyle practices (norms of behaviour) and sustainability issues and embed positive sustainability experiences in others. In theory, this ultimately results in long-term behavioural change, illustrating key elements in the behavioural change process from Lewin's model [147] of *'habit deconstruction'* or *'unfreezing'* challenging barriers that inhibit change followed by *'change'* when alternatives are considered, then reinforcement of the new procedures or *'re-freezing.'*

Communications in the form of *'eco-information'* should be tailored to the specific needs of the individual participant or to the organisation (generic resources have a lesser impact) in a localised and relevant way and should be positive and practical (Figs. 5.31 and 5.49). Facilitators should avoid criticism of opinions or actions and therefore guilt about existing unsustainable behaviour but should instead foster positive feelings that participants are going through a process which will help them make changes (eco-psychological transformation). By measuring and experiencing *'eco-feedback'* at first hand (Fig. 5.36) the environmental impacts of behaviours are easily understood and act as a continual source of motivation which fosters a sense of making a difference alongside other members of their team or peer group. Collaborative and integrative working is a key factor in behaviour change.

It is important to recognise that this is a process that requires continued support over time in order to ensure not only that change occurs, but behaviour changes are likely to endure. The Freudian theory of behaviour change emphasises the sense of self in relation to the environment and changes in behaviour that affect progress toward sustainability can also affect how people view themselves and their organisations. When people engage in actions that reduce CO_2 emissions, they are likely to see themselves as the type of person who cares about climate change based on their engagement with the behaviour. These changes in how they perceive themselves can significantly affect their support for process and policy changes.

However, the Norm Activation Model (NAM) of behavioural change highlights that the awareness and belief in positive consequences of sustainable behaviour are most effective for low behavioural costs (time, money and effort). These were cited by site managers as key barriers to on-site training of operatives and therefore represented high behavioural costs and tended to result in negative perceptions of the consequences of sustainable behaviour. From literature review and existing research [266] it was discovered that the traditional approaches to sustainable skills development in the UK construction industry are poorly structured, inconsistent, perceived with scepticism by the industry and use a vertical methodology where the same disciplines teach each other, e.g. architects train architects, etc. This mono-disciplinary approach has proved to be ineffectual in achieving significant behavioural change for the wider industry.

The case studies demonstrate that traditional construction materials and methods can work in harmony with innovative modern materials and techniques to create resource-efficient, *'climate-proofed'* contemporary buildings and that behavioural change theory recognises that change happens when it is relatively easy and convenient to assimilate that change, demonstrated by the integration of 'natural' materials into mainstream new-build projects and the integration of energy efficiency measures into traditional buildings. A number of the case study projects struggled with finding suitable measures of strength and quality for many new or innovative sustainable materials, illustrated by the experience with the WISE building rammed earth wall. There appears to be a gap in the British Standards which many of the projects had to deal with as they arose highlighting a need for further consolidation of standards and guidelines for sustainable materials. This could have a significant impact on the increased use of sustainable materials in the mainstream construction industry representing 'eco-choice.'

Evidence from interview data analysis shows that contractors perceptions of some sustainable materials had changed through direct experiential learning leading to a stated change in future behaviour to the use of more sustainable materials, such as low environmental impact paints and finishes. One respondent stated that 'the lesson is you don't have to have expensive materials for them to look modern and interesting and still achieve high levels of sustainability.' This comment reinforces the role a building can play in raising awareness of sustainability while dispelling certain prejudices about sustainable buildings which challenges *'norms of acceptance'* and has a *'cognitive'* influence in completing missing or imperfect information. This creates behavioural change opportunities by making explicit what is feasible and instils attitudes and perceptions that show it is not so difficult or inconvenient to act or change behaviour to achieve more sustainable outcomes.

Conversely, negative experiences correlated with stated rejections of certain practices that impacted the building schedule and can be seen as a justification for off-site training, away from the working environment, which is often seen as sacrosanct in the building industry. This allows learning to take place in a more psychologically safe environment where mistakes can be made without incurring high behavioural costs for the project in terms of time, money and effort. This also serves to increase positive perceptions once the learning has taken place. Observations and responses from site workers indicated that on site innovation does not work well and acts against the likelihood of behaviour change.

It was noted that natural materials are inherently degradable, when compared to man-made materials despite them having greater sustainability in terms of recyclability, embodied energy and low toxicity. It was stated that there needs to be a change in behaviour and attitudes in how we build and maintain sustainable buildings according to the materials and techniques used. Many sustainable materials require better workmanship and detailing. As one interviewee put it 'If the building is loved (a psycho-emotional connection) it will be maintained resulting in longevity, durability and greater sustainability.'

It was widely acknowledged from findings from the online survey, interviews and supported by the literature review that traditional, well-established and well-understood construction methods that also have sustainable benefits (given the current low level of sustainability skills in the construction industry) tend to work best. These include timber framing (Fig. 5.3) or using sustainable materials in a traditional way until more sustainable practices become the norm. The findings have shown that some skills can cross-over trades and skill sets such as the timber structures erected by steel erectors and the hemp-lime mix applied using existing concrete sprayers (Fig. 5.5).

Behavioural change theory, as discussed in Chap. 3, shows that ease and convenience of assimilation between traditional methods, materials and technologies with sustainable options is a reliable way to bring about behavioural change to bridge what has been variously referred to as the 'value-action gap,' the *'intention-behaviour gap'* or the *'credibility gap.'* This acknowledges the schism, real or perceived, between a desire for sustainability and actually achieving it. Bigger challenges arise when actions or behaviours are seen to be outside the realms of *'norms of acceptance,'* whether they be social norms in terms of going against conventional wisdom or obligation, technical norms in terms of orthodox approaches to labour and materials, regulatory norms, in terms of laws, rules and standards and cognitive norms, in terms of prevailing cultural beliefs or where missing or imperfect information prevents action.

Each of the buildings offer *'eco-information'* on sustainable construction materials methods and technologies (Fig. 5.86) and give the opportunity to view and experience high-quality buildings which have been constructed using sustainable methods. They enable the exploration, explanation and evaluation of cutting-edge thinking in sustainable construction by introducing the use of sustainable practices and materials into the mainstream construction industry. Some of the projects have strong links with the construction industry and built environment professionals can visit for new and innovative ideas, to find out how sustainable construction works in practice.

This is important in challenging behavioural norms and the buildings can operate on a cognitive level by filling in missing or imperfect information, acting as an *'eco-spur'* by encouraging practitioners to adopt resource-efficient working practices embedded in the building and experiencing *'eco-technology'* that ensures behavioural change can take place by providing the hardware to enable change. Habit deconstruction can then begin to take place by challenging barriers that inhibit change (un-freezing), considering and adopting alternatives (behavioural change) and set in place new procedures to prevent reverting back to old habits (re-freezing).

A whole-building approach moves away from linear working relationships and encourages a greater sustainability literacy to develop and challenge the conventional knowledge base (norms of behaviour). Sustainability requires the reconfiguring of standard operating procedures and was shown, particularly during the construction phase, to invite considerable resistance across the five case study buildings. The organisational and institutional resistance was shown to be driven largely by social influencers such as conventional wisdom, uncertainty of policy and regulation and technological constraints such as raw materials, labour and energy which are difficult to quantify beyond capital costs and it was found that lifecycle and payback are rarely factored-in, as a result. Mono-disciplinary working practices resulted in abortive work and project inefficiencies and a general lack of awareness of other disciplines concerns, exemplified by the failure of the rammed-earth wall on the WISE structure resulting from a resistance to engage with specialists.

Research from this study (see Sect. 2.6) has shown that directly relating environmental performance to financial costs can have an impact on increased sustainable behaviour. One suggestion elicited from the interview data analysis was to charge building users and visitors for the resources used during their occupancy. This would serve as a tangible and direct personal experience of their economic and environmental impact with the potential for behavioural change.

The function and ethos behind the organisations can be reflected through the buildings themselves with both educational and environmental aims influencing the behaviour of users and visitors. The users of the buildings are more likely to have had input at design stage under these criteria and therefore the design intent is more likely to be realised, reducing the value-action gap.

A significant impact of the buildings beyond the site itself was to change embedded (norm-centred) organisational behaviours, such as the procurement of the copper roofing system for The Core building (Fig. 5.39). This led to certified and sustainably sourced copper products being more easily and readily available for the global construction industry, illustrating significant sustainable innovation and organisational behavioural change, as a result of direct experience with an exemplar sustainable building system. The process makes it easier and more convenient for the industry to adopt this 'eco-choice', recognised in behaviour change theory as two key factors in enabling behaviour change.

Generally, respondents to the on-line and interview surveys showed an overwhelmingly positive attitude when asked about sustainable features of the buildings, particularly related to the overall sustainability of the architectural design, energy and water use followed by recycling and waste, transportation and indoor environmental quality. Changes in attitude and behaviour towards the sustainability of materials was positive but with some negative connotations. This could largely be explained by the use of innovative construction methods related to sustainable materials that were either not understood, poorly communicated or there was considerable scepticism about their use through lack of experience or pre-existing prejudices.

This proved to be the case for some of the projects where innovative materials and related construction methods failed because of poor on-site practices and at post-construction stage did not perform as expected. An observed example of this was the rammed earth walling for the rotunda lecture theatre (Fig. 5.4) as part of the WISE complex, a signature feature which has shown favourable responses in terms of aesthetics and sustainability but poor responses in terms of *'buildability'* from the contractors and poor acoustics from users of the lecture theatre.

This research has found that involvement of all key stakeholders throughout the design, construction, operation and use of buildings is recommended through an integrative, multi-disciplinary approach in order to ultimately establish a holistic understanding and awareness of the buildings in occupation. There is a strong correlation between sustainable behaviour in-use and the buildings environmental credentials for optimising the overall performance of a building, and therefore a powerful argument for the participation of building users and operators at design and construction phases.

Across all the buildings investigated for case study from the online survey the use of sustainable materials has performed best as a key feature encouraging sustainable behaviour, with a significant pedagogical impact on students and visitors, followed by energy efficiency, renewable energy and environmental management. This correlates well with previous findings that show these factors to be readily incorporated into sustainable buildings and used as teaching resources while the lowest scores from the responses show features that are harder to communicate via the building itself, such as sustainable transport, biodiversity and water conservation which tend to be either related to the geographical location or reliant on infrastructure. Some of the less tangible criteria such as health and well-being and pollution do score quite highly and this is reflected in some of the textual responses for this question.

Some of the most successful measures found to have the greatest psychological, physiological and pedagogical impacts resulting from literature review and supported by findings in this research and discussed throughout this thesis are the incorporation of natural day lighting (Figs. 5.11 and 5.12) which has been shown to improve cognitive function in educational establishments and productivity in the workplace and the use of natural materials which create healthier indoor environments, have low environmental impact and serve a strong aesthetic function.

Providing building performance feedback also rated highly serving to raise awareness of the impacts of user behaviour in relation to the environment. An appreciation of the influence of nature in the design, construction and operation of the buildings (biomimicry) also had a strong psycho-emotional and physiological impact (Figs. 5.52 and 5.60). Biomimicry can create innovative, efficient and environmental architectural solutions, making sustainable features accessible and visible to users along with clear and concise explanation and interpretation giving opportunities for individual and group experiential learning, shown to have a relatively high cognitive impact compared to other behavioural change strategies.

When asked *'In your opinion what key features of this building have had the greatest impact (positive or negative) in terms of environmental awareness and sustainable behaviour?'* the key responses included that the buildings were more *'natural and not mechanical'* reflecting the philosophical and theoretical discussion in Chap. 2 around intelligent buildings versus interactive buildings; *'information provision and transparency of methods, technologies, structure and materials'* reflecting the pedagogical efficacy of the buildings; *'promotes the association of eco-design with thoughtful and good, modern design. Eco does not mean bad design'* reflecting the aesthetic principles discussed in Chap. 1; *'the combination of sustainable features, working together'* showing how congruency helps people order information and develop new norms of acceptance; *'positive experience that people wish to replicate.'* Positive experiences help to reinforce new norms of behaviour that can then be repeated and *'the environmentally sensitive use of materials'* again makes the sustainable methods, materials and technologies more tangible and adoptable. The most negative experiences were related to *'proximity and car reliance'* which can reinforce perceptions that sustainable buildings are remote, purely rural ventures.

When asked *'do you feel there is anything that could be done differently at post-construction stage to increase the level of sustainable behaviour of the occupants of the building?* key responses included; *'transparency,'* again reinforcing the need for the buildings to be open, accessible, intelligible and coherent; *'5–10 min explanatory tour of the building for all visitors,'* a cognitive strategy filling in missing information, *'more access to the building generally, services, specifically,'* for educational purposes *'more feedback,'* showing a need to understand the impact of behaviour and user guidance for visitors.

One survey question elicited responses as to whether any building user guidance had been provided in relation to the sustainable and efficient operation of the buildings in order to establish what has become known as the *'soft landings'* approach, as discussed in Chap. 4. This encourages the design and construction professionals to engage more with the operation and use of the building, particularly in the early post-construction phase, with operational information and guidance in order to optimise the environmental performance of the building as well as the awareness and sustainable behaviour of the occupants, including staff, students and visitors.

Generally, from the responses given it can be concluded that user guidance and training is considered highly relevant, when it is provided but can be sporadic and in some cases confusing. It is clear that early intervention is critical in establishing beneficial habitual behaviour that has implications for long-term performance of buildings and sustainable actions of occupants. In terms of using the building as an educational resource, knowledge and understanding of the sustainable features of a building is critical, in order to develop teaching and learning activities based on those features.

If only certain stakeholders are considered for training in the use and operation of a building, e.g. facilities managers, engineers and caretakers, while others are not, it may reasonably be argued that the majority of occupants may not be using or operating the building to its optimum performance level in terms of technical functions related to heating, cooling, lighting, ventilation, equipment, etc., but also in terms of the pedagogical opportunities for using the building as a teaching and learning resource.

Generally, it was found that there is a lack of culture of learning from buildings in use, highlighted by findings from the case study buildings of minimal focus on post-occupancy performance, despite their sustainability and pedagogical credentials. Literature review shows that there is a significant need to assess buildings during occupancy. Post-occupancy evaluation of buildings has the potential to focus the minds of designers and contractors at the beginning of the project considering both technical and behavioural factors in the performance of a building, particularly in response to user needs, adaptation and design quality, creating a virtuous circle of improvement.

Post-occupancy data can help find solutions to the performance gap and can be used as a behavioural tool to provide performance data to building occupants and users, so long as this is done in a user-friendly way. Traditionally user interfaces have been difficult to use but digital media, photo realism and the use of 'apps' has made interaction much easier and is a field of continuing development.

Negative experiences outside social norms of accepted behaviour (conventional wisdom) in working practices can have unconstructive consequences for the learning process and behaviour change, reinforcing prejudices against new methods, materials or technologies. Barriers can be *'cognitive'* in terms of missing information and *'behavioural'* in terms of obstructing sustainable actions by making choices difficult and inconvenient. This research has shown that psycho-emotional blocks and barriers can be overcome through training and 'learning by doing,' awareness raising and a degree of compulsion through regulatory and statutory requirements to both motivate and reinforce sustainable behaviours, as discussed in Chaps. 3 and 4.

One of the interviewees stated that, 'there has been a slow and incremental change in the people who are touched by the building towards environmental behaviour, immediate conversions (to sustainable behaviour) are not expected or achievable.' This is supported by the work of Global Action Plan whose behavioural change strategies recognise the need for incremental change over time.

6.7 Other Stakeholders and Behavioural Change

The online survey clearly shows that attitudes vary considerably between stakeholder roles in their willingness to engage in the consideration of a buildings impact on personal behaviour. The relative psychological and time-related distancing of design teams from the use and operation of a building is recognised as a key barrier to sustainable buildings in achieving both technical and behavioural sustainability goals, known widely as the 'credibility gap' and is supported by the 'soft landings' approach developed by Leaman and Bordass [149], as discussed in Chap. 4, which proposes greater responsibility of design and construction teams for the performance and operation of their buildings requiring the identification of behavioural goals and a behavioural commitment to achieving them.

The word cloud in this chapter quantifies the perceptions of the building by the respondents overall which show that *'light'* and *'airy'* have a significant number of responses at 12 and 4.7% respectively. This corresponds well with recognised research and is supported by the findings from the extensive literature review into day lighting and natural ventilation as two of the key attributes of sustainable buildings with both physiological and psychological implications for users, as discussed in Chap. 3 and illustrated in Chap. 5 (Figs. 5.2 and 5.81).

Other significant physiological factors from literature review that correlated with the primary research findings from case study analysis were found to be thermal comfort, indoor air quality, impact of materials and acoustics which relate closely with psychological factors of well-being, feelings of calmness and tranquillity, aesthetics, inspiration and an awareness of the sustainability and educational worth of the building.

Some of the more physiologically perceived responses relate to temperature: *'cold'* (2.7%) and *'warm'* (1.3%), indoor air quality: *'stuffy'* (0.7%) and *'smell'* and *'noise.'*

The word *'sustainable'* with a frequency response of 8% is significant in that it shows a relatively high level of awareness of the broad purpose and function of the building. Other responses associated with the environmental credentials of the buildings are *'natural'* at 4.7% with *'green'* and *'eco'* both at 2.7%.

Some respondents did not like the terminology used around sustainability such as the use of the words *'eco'* or *'green.'* It was felt that these gave a negative or less serious connotation to people outside the sustainability community. One project leader did not want to term everything as *'eco'* so as not to alienate people who do not associate themselves with the environmental cause, outside social norms of acceptance. One interviewee noted that his reserve in using such terms in his courses and project materials is actually 'shooting him in the foot' because the government is now sponsoring the use of the term green with initiatives such as the Green Deal.

The words *'beautiful'* (3.3%), *'calmness'* (2%), *'welcoming'* (1.3%) and *'comforting'* (0.7%) arose as key perceptions from the online survey and interviews. They associate positive psycho-emotional connections to the building. It was discovered from primary and secondary research findings that psycho-emotional responses to phenomena are a critical factor in encouraging, or indeed discouraging, behavioural change. Psychological blocks were evident with several respondents citing the 'macho' culture of the construction industry where it was often perceived as socially unacceptable (the cultural norm) to engage with sustainability issues. Theories of behavioural change show that a closer personal connection to a problem can elicit emotional responses by making the threat tangible and establishes an easily understood causality. The study of the behavioural change work of Global Action Plan reveals that environmental pressures and emotional fears and concerns about climate change and global warming should be used to encourage people to take practical action in social groups.

The Gestalt theory, described in Chap. 3, supports this assertion and recognises humans innate ecological wisdom and personal responsibility for the environment within a social context where people begin to appreciate that our behaviour and environment mutually effect each other, e.g. consumption patterns have a reciprocal relationship with energy costs. Some of the mechanisms that inhibit sustainable behaviour, despite efforts to be more sustainable, include *'moral-self-licensing'* where people feel that doing something good, e.g. super-insulating a building allows us to do something bad, such as have the heating on constantly, this was also highlighted as the *'rebound effect.'*

Emotional triggers are commonly used in traditional marketing to persuade people to buy products and services, to change consumer behaviours, particularly in financial, health and leisure industries appealing to strong psychological and emotional feelings. Social marketing uses these strategies for social good using positive reinforcement, indirect suggestions and targeted behavioural messaging. The case study buildings, to varying degrees, display aspects of these emotional responses. From the interview and survey data analysis, emotional responses to the buildings, their materiality, function and ethos were evident from many of the respondent's answers as presented above and from the repetition of the expression 'the building has the "*WOW*" factor' (Fig. 5.9).

Responses such as *'inspirational'* at 3.3%, *'educational'* at 2%, *'thought-provoking'* and *'exemplar'* suggest that respondents feel the buildings have a pedagogical effect on both themselves and other building users. Some responses were more scientific including *'technically challenging,' 'efficient,' 'innovative'* and *'robust'* and some were directly related to the elemental materials used such as *'wood'* (3.3%), *'earth'* (0.7%) and *'stone,'* key sustainable materials that are directly associated with low environmental impact, natural and healthy attributes (Figs. 5.16 and 5.40).

Some responses had negative connotations; *'cold,' 'leaky,' 'sterile,' 'unpredictable,' 'noisy'* and *'frustrating.'* This illustrates both the variability in human physiology and perceptions of comfort. The words *'unpredictable'* and *'frustrating,'* were identified as coming from building and facilities managers and highlight that even the most sustainable of buildings will have common problems in their operation and use as well as specific issues because they are adopting often unfamiliar, untested and innovative sustainable practices and techniques. This can have a significant impact on both the sustainable behaviour of individuals and the sustainable performance of the building. This was amply illustrated by findings from the CORE project where the unfamiliarity of using a building energy management system, and lack of guidance, combined with technical difficulties lead to its circumvention. This was recognised by the operative as sub-optimal and undesirable.

The five case studies show that linking environmentally sound buildings with mainstream business and educational practices are possible and offer a location for people to explore practical sustainable solutions. The buildings embody sustainability principles that are encouraged and taught within them, offering experiential examples of sustainable technologies and lifestyles. They offer a tangible demonstration of state-of-the-art sustainable design and a glimpse into the future practice of building design linked with sustainability and education. Current and future built environment professionals are able to test out the building's special features on tours and in workshops and classes taught by experts in sustainability.

Each of the case study buildings uses displays, exhibits, signs, prompts, modelling, demonstrations and information provision to varying degrees as antecedent interventions and mechanisms to encourage and enable behaviour change involving experiential learning through self-directed and user-centred inquiry and play, shown to be among the most effective ways of instilling information and knowledge, as illustrated by the learning pyramid in Chap. 3 (Fig. 3.11). This also fulfils the cognitive element of the behavioural change process which highlights that unsustainable behaviour arises from missing or imperfect information.

A number of respondents noted that some people find it hard to take the sustainable lessons back to their *'smaller environment'* such as their office or home when they learn about them in a large centre completely dedicated to sustainability. This sense of powerlessness or being overwhelmed by a problem relates closely to pre-existing cognitive and social norms of behaviour and can be attributed in part to the 'passive-bystander effect' which states that people tend to exaggerate their own powerlessness. This can be tackled by designers and educationalists by making sustainable features, interventions, materials, methods and technologies relevant and applicable at various scales and across a number of audience profiles. For example, rainwater harvesting can be applied on a large industrial scale but the principles can function equally well on a small domestic level (Fig. 5.35).

From the theoretical study it was found that human behaviour is generally nonlinear and non-rational and people tend to be *'boundedly rational'* meaning they behave rationally within certain limitations. The interview and survey data analysis shows that despite being involved with an exemplar sustainable building many people still behaved in unsustainable ways, such as opening windows for cooling when the heating is on, wasting materials that could easily be utilised, leaving materials exposed to the elements, leaving electrical equipment on when not in use, etc.

This demonstrates a number of psychological and behavioural change mechanisms that inhibit optimal sustainable behaviours. As already identified *'moral-self licencing'* or contradictory behaviour reflects the limitations of behaving sustainably when we feel that doing something good allows us to do something bad. This was shown to be the case by a number of responses from site workers, where working on a sustainable building was perceived as being *'good enough'* and offset the need to consider the sustainability of their own working practices.

This is similar to the *'rebound effect'* whereby an improvement in sustainability can lead to an increase in consumption. Evidence for this effect was shown by a number of the case study buildings where the effectiveness of the buildings, as a sustainability and educational resource for sustainability, was offset by their remote geographical location, poor public transport infrastructure and the need to rely on car use.

A significant number of staff within the buildings stated from the written responses for the online survey that their behaviour had been impacted through working in a more sustainable environment in terms of physiological and psychological health and well-being with a direct impact on their ability to focus on tasks resulting in improved productivity. This finding is supported by the Usable Building Trust's Building User Survey for the WISE complex which showed perceived increased productivity of between 5 and 15% by 62% of respondents.

The research findings presented in this chapter suggest general levels of behavioural impact across the different user groups with higher levels of impact shown on teaching staff, students and visitors, followed by general staff, maintenance staff and community groups. This would indicate a greater need for the participation and involvement of the latter groups in the general environmental and sustainable attributes of the buildings.

The efficacy of interventions with an explicit and implicit attempt to change behaviour by the five case studies is variable but there is clear evidence in a number of instances where behaviour has been changed through engagement with the building project. From the findings contained in this chapter, these broadly include clients and design teams changing their organisational practices, contractors learning about and adopting new sustainable construction skills, methods, materials and technologies, the improvement of staff productivity and awareness of their environmental impact as a result of working in a more sustainable and healthy building and visitors and course participants being encouraged to learn about sustainable buildings which has had a stated effect on their lifestyles, with the potential for adopting more sustainable practices through formal, informal and experiential teaching and learning experiences.

The coding and reporting of the interview transcripts have illustrated the integral relationship between people and buildings and more specifically between the pedagogical potential of each of the five case study buildings and the sustainable methods, materials and technologies embedded throughout them. From the interviews it is clear that in the sustainable design, construction and use of the case study buildings and the sustainable ethos of both individuals and organisations there is an explicit agenda to raise awareness, educate and ultimately change behaviour which is rarely evident in the majority of buildings.

If this intent was adopted more widely it could have a significant impact on economic, social and environmental sustainability throughout the built environment with repercussions for carbon reduction and less reliance on finite resources. One aspect of the behavioural change potential of buildings was seen to be subliminal, that behaviour change may not be the reason for visits but is a consequence of experiencing the building (experiential learning).

6.8 Pedagogy and Behavioural Change

The buildings lead by example on a range of issues relating to climate change and sustainable development. The buildings themselves are important in promoting sustainability to learners, community groups, organisations and the general public. Many are intended to be a catalyst for the development of an innovative curriculum, filling the identified cognitive gap for key skills and knowledge.

As a very visible and tangible showcase of best practice in sustainable design, construction and operation, the buildings are a learning resource in their own right and, in the words of one respondent to the interviews, the building 'motivates users to put their new skills and learning into practice.' This reflects the idea of the building as pedagogy, as third teacher and part of the *'hidden curriculum,'* as discussed in Chap. 3. Each of the projects showed a level of social responsibility to prepare students and young people for future climate changes that could happen in their lifetime and reflects the recognition of a social obligation on organisations for shifting social norms.

Particular features of the buildings that encouraged teaching and learning were identified from the interview data analysis. These included a discussion around the lifecycle of materials, energy efficiency and energy policy, use of natural daylight and passive solar design leading to discussions on the greenhouse effect and global warming, thermal mass, heating and cooling and renewable energies.

The buildings display a strong congruency and correlation between their form (aesthetics and exteriority) and their function (meaning and interiority), departing from the idea of the building as a pure artistic expression which encourages architectural aesthetic obsolescence or buildings reflecting fads or transient trend. Translating this profound and philosophical aspect of sustainable buildings, allied to their lifecycle and historical context offers a coherent narrative which appeals to our inherent need for order and meaning and can be utilised for educational purposes. One example, the heat recovery from the waterless urinals at the WISE complex, was cited, illustrates the narrative by bringing to life the idea of an everyday human function that everyone can relate to which can have positive sustainable consequences through the application of sustainable technology.

The building itself is also offered as an interactive exhibit and to be a *'test bed'* for public education leading to behaviour change. Interactivity is a key factor in experiential learning which, as shown in the learning pyramid (Fig. 3.11) encourages and enables knowledge retention and a deeper, third order and transformational learning experience involving redesigning and enabling an alternative world view and a shift from *'transmissive'* pedagogical techniques involving imposed instruction to *'transformative,' 'participative,'* and *'constructive'* techniques [185].

It has been shown that the key drivers for the buildings are pedagogical, environmental and in some cases theological, as in the Quaker principles embedded in the Sidwell school project, which includes a dedication to environmental stewardship. From the outset it was a belief that a sustainable building provides an opportunity to achieve an outstanding level of integration between the curriculum, values and the mission of the school. The school's sustainable features are intended to teach and inspire current pupils and for those students to carry and transfer their knowledge and appreciation of natural systems into the future.

The involvement of those engaged with the functioning and purpose of the building is critical at design stage and is particularly true of buildings with a pedagogical agenda where the building can be used as a teaching and learning resource. The sustainability ethos is often driven by an individual sustainability champion, often a teacher, who acts to maintain the sustainability credentials of the project and the involvement of educators was particularly evident in each of the case study buildings, who arguably have a greater understanding of the potential for behavioural change through experiential teaching and learning strategies, and in all probability have undergone some degree of eco-psychological transformation from their own training and experiences.

From the online survey respondents reported that the biggest positive impact of the buildings (and therefore potential for sustainable behavioural change) was noted for students and visitors to the buildings who were informed of the sustainable credentials, the buildings used as an educational resource, with the aim that they would be motivated to live more sustainably by experiencing sustainability in practice and some teaching staff noted a reciprocal benefit from receiving questions and feedback from students and visitors challenging them to consider the building and its sustainable credentials at a deeper level.

This type of teaching and learning reflected in the case study buildings and discussed in Chap. 3, based on the writings of Professor Sterling [185] illustrates education *for* and *as* sustainability, a transformative approach which is constructive and participative as opposed to instructive and imposed. It also reflects the notion of second- and third-order educational change. The former enables critical reflective learning and some reformation of the existing paradigm to reflect more thoroughly the ideas of sustainability and the latter facilitates a transformative learning experience as a creative and reflexive process.

The shift is towards *'learning as change'* which engages the whole person and the whole learning institution. Sterling highlights this response as the most difficult to achieve, as it is most in conflict with existing structures, with respect to unqualified economic growth, for example. This gives an indication why education *about* sustainability with a content/knowledge bias leaves basic beliefs unchanged, being readily assimilated into the existing educational paradigm, and represents first-order change. Essentially, educational institutions need to undergo a process of change and transformation in order to facilitate change and be transformative themselves. Each of the case study projects represent organisations that aspire to second- and third-order change.

As each of the buildings selected for case study have both a strong pedagogical and environmental agenda a number of respondents indicated that evaluation of the building's performance in terms of resource efficiency, indoor environmental quality and user behaviour was included as part of the curriculum and therefore, almost by default, significantly raised the awareness of how the building was performing with concurrent behavioural effects because of that increased awareness. This was highlighted by the follow-up survey sometime after the completion of a number of the buildings, which suggests time is a critical factor for curriculum developers to assimilate the environmental features of the building into their teaching and learning strategies.

Education for sustainable development (ESD) modules and courses, related to the built environment have increased in popularity, being delivered at many levels of formal education, and professional and vocational training, particularly evident in the rising demand for places on CAT courses, a key driver for the production of the WISE building.

ESD related to the built environment is increasingly being delivered in schools of the built environment and as a requirement for continuing professional development (CPD) but remains a minority topic for all but the most enlightened educational institutions. There is a strong argument for teaching sustainability across curricula and across disciplines to offer a holistic approach to sustainability embedding and relating it to all aspects of human lifestyles as well as the built environment.

The online and interview surveys asked 'are there any other features of the building you feel would enhance teaching and learning about the environment or significantly encourage sustainable behaviour? Key responses included: *'more visual information,' 'signage,' 'interpretation'* and *'detailed data on performance of systems'* which again highlight the pedagogical requirements of many of the building users to have an impact on other users and visitors to the buildings. *'The overall feel and philosophy of the building'* was cited as an influential factor on teaching and learning. This is a less tangible, more complex and harder to define aspect of the buildings and therefore more difficult to quantify but was a significant and influential factor on the perceptions, activities and behaviours related to the buildings.

Other responses included; *'greater transparency, particularly EMS, energy, water and resource use'* reflecting the need for the building itself to be made accessible for investigation for educational purposes requiring pedagogical factors to be taken into account at design stage with the involvement of educators. Negative factors were *'noise transmission'* a result of the use and misunderstanding of the implications of new and innovative sustainable materials methods related to the function of the building, most notably the WISE rammed-earth lecture theatre which reverberated sound and the open plan DACE workshops which allowed the noise from conflicting activities, seminars and workshops to interfere with each other.

6.9 Chapter Summary

This chapter has drawn together findings from the research and shows that by considering whole-life building strategies the built environment has the potential to be a valuable catalyst for pro-environmental and sustainable behavioural change among built environment professionals, building managers, occupiers, teachers, staff and visitors.

Buildings can and should be used as a teaching and learning resource for developing an understanding of sustainable methods, materials, technologies and behaviours over the whole of their lifecycle. Ideally all buildings, but especially those purporting to be *sustainable and pedagogical* should explicitly and implicitly influence sustainable behaviour in their design philosophy, construction processes and operational practices. The following concluding chapter summarises how the research objectives have been met, the original contribution to knowledge, limitations, deficiencies, future research in this field and ends with a summary of key findings and recommendations.

Chapter 7
Conclusions and Recommendations

This final chapter presents the conclusions drawn from the main findings of research based on the theory and practice of sustainable construction and the promotion of sustainable behaviour.

7.1 The Need for Sustainable Buildings

The environmental impact of the built environment was set out and its potential for reducing carbon emissions thus contributing to offsetting anthropogenic climate change through technological and behavioural strategies with additional physiological, economic and social benefits. The impacts of global warming through climate change will have implications for the planning design, construction operation and use of buildings, towns and cities which will need to adapt in order to mitigate risks by significantly reducing the burning of fossil fuels and designing for zero carbon buildings.

Technical and economic metrics in the built environment industry are well-developed and understood. Environmental metrics are less developed but are becoming mainstream through environmental rating systems and greater awareness throughout the building industry of environmental responsibility through education and legislation.

Social, psychological and cultural metrics are underdeveloped in the built environment industry and social sustainability is external to common decision-making and requires further investigation and application. Developing a sustainable and user-centred theory so that it can be applied to a wider range of building typologies and projects will increase environmental support for building activities. Applying knowledge of the users into the supply chain will eventually change the way the industry operates. In order to overcome social and psychological barriers transformations in

J. L. Clarke, *Sustainable Buildings for Sustainable Behaviour: Theory and Practice*, Green Energy and Technology, https://doi.org/10.1007/978-3-031-98905-6_7

the unconscious value systems and the habits and norms of construction organisations are needed.

It is also shown throughout this study that buildings have a profound effect on the physiology of human beings; their health, well-being and performance, identified as strong social drivers for the provision of sustainable buildings. As awareness of the behavioural, cognitive and eco-psychological effect of sustainable buildings increases, the natural human response is to better understand the physiological benefits of sustainable buildings by encouraging and demanding design for sustainability with increased benefits of natural daylight, ventilation and energy efficiency.

Greater public exposure to, and experience of, exemplar sustainable buildings will help promote their features and it may be argued that intelligent critiques will help facilitate their acceptance. A more realistic expectation about their intended and achieved environmental, economic and social performance will help inform and improve future projects. Sustainable building could become so much of a standard practice that sustainable building will become mainstream and therefore the term sustainable building will become obsolete.

As presented throughout this book this will require more than just a development of sustainable technologies and lower costs for these technologies. By identifying social and psychological barriers, social structures, rewards and incentives can influence change which can be incorporated into every decision of the building process. This transformation cannot happen without structural changes in organisational systems, concurrent with adjustments to society's unconscious value system.

Buildings can enhance the natural environment by encouraging biodiversity and the sustainable use of resources. Sustainable buildings can be used to raise our awareness of the finite nature of resources through energy efficiency, water conservation and how we process and dispose of waste materials.

Economic metrics in the built environment are well-defined and internalised in decision-making. Decisions to build in most modern societies are still guided by conventional business drivers, which include market conditions, financial profit, construction processes, technology and competitive advantage. The current mainstream building industry assumes the sustainability question can be resolved through improved construction technology, increased factor productivity (material efficiency) and market forces alone (false positivity).

The fear is that policies effectively mitigating ecological degradation would slow economic growth. Given advanced scientific evidence about global warming, presented in this study, it is clear that this argument is weak, as the consequences of climate change and ecosystems collapse will not only slow growth, but could also destroy the global economy. Green growth, responsible capitalism and capitalism with a difference are proposed as better future economic practice.

The most effective driver for a more sustainable architecture by far would be if environmental impacts were reflected in real costing, including environmental costing of materials and energy. This will only happen in a market economy if the costs and benefits to the world are reflected in the costs and benefits to individuals in line with the global context. To achieve this, the bottom-up transformation without coercion through cultural change will almost certainly need to operate in parallel with a top-down process of policy change driven by international obligations. The emerging sustainability market is seen by many as an entrepreneurial opportunity to meet the increasing demand for sustainable buildings.

Sustainable buildings are the embodiment of the identification of environmental problems (in built form) and can be a powerful behaviour change agent for future practice and new norms of social, economic and environmental behaviour.

7.2 A Clear Definition

Broadly, sustainable building may be defined as building design and construction that uses methods, technologies, materials and processes that are resource efficient and that will not compromise the health of the environment or the associated health and well-being of the building's occupants, construction workers, the general public or future generations [18].

According to the European Economic and Social Committee (EESC) glossary of sustainability in the built environment [19] sustainable construction is defined as 'the application of sustainable development principles to the design and construction process i.e. use of fewer virgin materials, less energy in construction, less energy in use, less pollution and less waste; *whole life* approach to design, construction and life use; and providing safe places and work with acceptable social conditions integrated into sustainable communities.'

There are a number of international design standards developed over recent decades that have reached a high level of complexity in defining and assessing the environmental impact of buildings. The UK Building Research Establishment Environmental Assessment Method (BREEAM) has been adopted for the purpose of this study as a model benchmark of sustainability as the basis for comparing and contrasting the selected case study buildings.

Environmental behaviour has been defined as 'behaviour that changes the availability of materials or energy from the environment or alters the structure and dynamics of ecosystems or the biosphere' [130], p. 12. Sustainable behaviour is used as a term to describe behaviour that has a positive influence on the local, national and global ecosystems or biosphere.

Brundtland [34] proposed a broad definition of sustainability, adapted for this study to reflect the behavioural element, as 'behaviour that results in the satisfaction of our needs today without diminishing the prospects of future generations to do the same.' This was refined by applying the triple bottom line concept of sustainable development resulting in the definition of sustainable behaviour for the purposes of this study as being 'the behaviour of individuals, organisations and institutions that achieve positive economic, social and environmental outcomes.

Pro-environmental behaviour is referred to throughout the study and differs from sustainable behaviour as it is a more specific term that focuses on purely environmental outcomes of behaviour which may not necessarily result in economic or social benefits.

7.3 Stakeholder Behaviours

This study has shown that decisions made by individuals as part of larger organisations throughout the whole life cycle of the building effect energy use, CO_2 production, pollution levels, waste and the use of finite materials and resources and those decisions strongly influence the behaviour of individuals and organisations. Working on and in a sustainable building should explicitly and implicitly influence the sustainable behaviour of its clients, designers, builders, owners and occupants which will have a reciprocal effect on the sustainability and environmental performance of the building. Therefore, it is argued that human behaviour needs to be considered from the outset of the process of developing a building and sustainable buildings should be designed, constructed and operated to encourage, enable and facilitate sustainable behaviour.

Sustainability in architecture emphasises long-term thinking and the role of the architect in enabling the sustainability of our environment and society represents a significant factor and a creative challenge. Over the years responsibility for performance of a building has shifted from the architect to the mechanical and service engineer. There is now enough peer pressure within the culture of architecture for architects to be uncomfortable with being associated with the more obvious examples of energy profligacy, material waste, high environmental impact and unhealthy buildings.

Architects and constructors rarely experience the results of their decisions on the performance of buildings and the 'soft landings' approach is highlighted as a future best practice approach.

Architectural education is traditionally focussed on studio design centres, isolated from the real-life collaboration reflected in modern construction projects. Architects have been criticised for becoming over-reliant on consultants, removing them from achieving quantitative performance targets. The educational establishment have been accused of producing 'semi-detached' professionals with little understanding of other disciplines in the process being perceived within the building industry as sole creators. Collaborative and integrative working from inception to completion and beyond is highlighted in this study as best future practice.

Universities and colleges have a key role to play in 'un-freezing' the teaching of unsustainable practices, changing behaviours and then consolidating sustainable behaviours through experiential learning in order to provide a sustainably literate future-built environment workforce including architects, engineers, constructors and building managers and sustainable behaviour needs to be applied, reinforced and consolidated through working practices.

Often, sustainability is seen as a threat to traditional working practices and conventional wisdom, in behavioural change theory terminology, social norms of acceptance. This was particularly evident in the attitudes of construction organisations. Generally, there is an essential requirement for individuals, organisations and institutions to fundamentally change attitudes and behaviours (habit deconstruction) in order to begin to achieve progress towards a more sustainable future.

This is strongly linked to the necessity for widespread education, skills and awareness raising in sustainable thinking at a fundamental level which needs to be addressed by governments in their provision of education for sustainable development, embedding sustainability at vocational and academic levels to achieve deeper, third-order transformational learning experiences. This applies strongly to the built environment sector which, as has been established, has a significant impact on the natural environment in terms of finite resource depletion and carbon emissions.

Typically, building users feature in theories of the built environment, along with the main players, but are rarely central to them. This is largely attributable to the difficulties of measuring human behaviour and the limitations of conventional social science research in the practical context of planning, designing, building, managing and occupying buildings. This study has investigated various theoretical perspectives related to sustainable behaviour.

This research has reported that contemporary systems of delivering the built environment have largely lost touch with the significance of the user, particularly in terms of the impact of their behaviour on the performance and operation of buildings. Since the industrial revolution the construction industry has developed an aversion to change adopting what are now largely considered unsustainable practices, procedures and technologies with an emphasis on cost rather than long-term sustainability and quality.

It is well-established that people often make decisions intuitively, effortlessly and with little conscious awareness. These choices are influenced by a mesh of environmental, social and economic factors. In some instances, the decision to build is driven purely by aesthetic considerations or building codes and standard specifications. As considerations for sustainability increasingly drive decisions to build, so the sustainable behaviours of stakeholders become a key factor and behavioural change can often result from engaging with the sustainability agenda.

Among all stakeholders in the provision and use of buildings there needs to be a common language of sustainability. In their design, construction and operation buildings need to avoid giving incongruent messages to users and allow them to easily engage with the buildings functions and operations to achieve low environmental impact and healthy environments for learning while understanding inherent sustainable features and how they can be realistically interpreted and made relevant to their own lifestyles.

This study has shown that decisions made by people throughout the whole life cycle of the building affect energy use, CO_2 production, pollution levels, waste and the use of finite materials and resources and those decisions have a reciprocal influence on the behaviour of individuals and organisations. The decisions of architects and engineers are important for setting the context for sustainable behavioural change and building managers and users' behaviour must respond in a sustainable way.

A sustainable building should explicitly and implicitly influence the pro-environmental behaviour of its designers, builders and occupants. Therefore, it is argued that human behaviour and behavioural goals need to be considered from the outset of the process of developing a building and sustainable buildings should be designed, constructed and operated to encourage, enable and facilitate sustainable behaviour.

7.4 Mechanisms of Behaviour Change

Both technical and human behavioural factors, particularly user behaviour, are key determinants in the performance of a building. How the buildings sustainable design aspirations, embodied in the varying aspirations of the stakeholders, are translated and embedded in the operation of the building, serve to reinforce and 're-freeze' new and sustainable behaviours.

The mechanisms that enable these changes have been highlighted as antecedent interventions such as education, signs, prompts, modelling, demonstrations and behavioural commitments and consequent interventions such as rewards, penalties and feedback. Each of these were demonstrated to varying degrees in the case study buildings and their effectiveness had differing effects dependent on the psychology, beliefs, attitudes and roles of a wide variety of individuals, as discussed in the previous chapter based on the theoretical principles and interview and survey data analysis.

Advances can be achieved through behavioural interventions, embedding processes, technologies and materials that enable and encourage sustainable behaviour such as providing building performance feedback, user control, natural ventilation and day lighting.

This study found that the potential for influencing the sustainable behaviour of building users is generally optimised at design stage with the involvement of contractors, operational staff and the users themselves through an integrative design strategy. Individuals and organisations must be open to the ethos of sustainability otherwise psychological blocks can hinder successful outcomes.

Some of the buildings that have been hailed as environmentally sustainable at the turn of the century are regarded as naive by some critics, whereas others herald them as exemplars. It can be strongly argued that we need bold but reasoned and justified design experiments to continue. The work of experimental architecture is important because it provides a series of reflective testing in practice that continues from project to project developing technical, environmental, social and economic innovation in sustainable development.

The study has highlighted deficiencies in some current working practices, in the knowledge and skills of built environment professionals which can ultimately be attributed to failings in their education and training. In many instances existing education and training do not embed sustainable design and construction skills to a sufficient degree of understanding in sustainable theory and practice to achieve the higher standards of design and workmanship required for sustainable building design and construction.

These deficiencies can result in negative perceptions being developed at an early stage from negative educational experiences or failures in practice. Over the years responsibility for the performance of a building has shifted from users to technological solutions. Building users and operators need to be assisted and encouraged to be conversant with how renewable technologies, passive and low carbon systems work in order to optimise the performance of their buildings. It is important to avoid unmanageable and over-complicated technological systems and scientific language to encourage user engagement and avoid users circumventing sustainable systems.

As the requirement for more sustainable buildings grows through legislation and consumer demand the lessons learned from the case study buildings in this research show that it is highly feasible to have sustainable buildings that can encourage sustainable behaviour and therefore many of the inter-related processes, technologies, strategies and materials can feasibly be incorporated into mainstream buildings to encourage sustainable behaviour.

Traditionally the environmental movement uses a rational and informing approach to behaviour change. This needs to shift to psycho-emotional triggers. This study has shown that merely raising peoples' awareness in terms of the sustainability of their own built environment and landscapes through information provision can have limited effects on long-term pro-environmental behaviour change, assuming that relating facts about, for example energy use will lead to better decisions.

This provides a firm rationale but leads to minimal impact. Telling people how they should behave has little effect on how they actually behave. Issues of willpower, motivation, cost and convenience are often more important than a lack of knowledge or missing information, known as the 'cognitive' factor in behaviour change theory. It has been found that sustainable building design makes sustainable issues tangible which people can relate to their own lifestyles.

In terms of the operation, use and the impact of buildings on sustainable behaviour there is strong evidence that sustainable buildings provide an effective and experiential learning environment encouraging deeper and more transformational learning as a basis for cognitive behavioural change. This enables people to ultimately make conscious decisions to act or change behaviours (eco-psychological transformational change) based on direct experiences of sustainable features and processes. It is a critical feature of behavioural change theory that people will have a propensity to change their lifestyles when it is easy and convenient to do so.

Behaviour change for sustainable development takes place through a complex series of socio-economic, socio-technical and psycho-emotional processes and requires conditions that are easy and convenient to enact, when there is a critical mass of early adopters and innovators and when the problem is perceived as a personal threat. Humans behave unpredictably and irrationally under certain circumstances (boundedly rational). They are highly adaptive and require responsive buildings with feedback without too much management dependency.

7.5 Causation or Correlation

This research has shown that sustainable buildings and sustainable behaviour are two comparable entities with a tendency to vary together indicating both the strength and direction of the relationship.

Therefore, it may be inferred that a strong correlation exists between those that engage in the design and construction of a sustainable building, as defined and individual attitudes and organisational practices. The correlation is evident throughout the lifecycle of a building and varies considerably between actors in the network.

However, 'correlation is not causation' and a direct causal link has not been proven between sustainable buildings and sustainable behaviour. Behaviour in relation to sustainable buildings, as defined has resulted in both sustainable and unsustainable behaviours across the diversity of stakeholders observed for this study and a broader and more longitudinal, quantitative study would be needed to establish a definitive and statistical causal link.

A significant amount of research in environmental psychology is based on how human behaviour is influenced by features of the spaces people occupy, that buildings cause users to behave in certain ways, some of which are predictable. Also, behaviour can result from learned social norms and patterns where buildings act at a symbolic level, as a mediator of the social relationships that determine behavioural outcomes. Building users' behaviour is also influenced by their feelings, intentions, attitudes and expectations as well as by the social context in which they are participating. It has been shown that limited time, energy and attention often lead to behaviours that are self-defeating or unsustainable.

Multiple sources of data indicate that attitudes formed form direct behavioural experiences are more predictive of later behavioural change than are passive or abstract attitudes. Therefore, it is concluded that the more we design, construct and use buildings as an experiential resource with pedagogical and behavioural change aspirations the more likely positive environmental behaviour is to occur. This is linked closely to the Freudian behaviour change theory of how we perceive ourselves in direct relation and proximity to the environment and nature. Our disconnection from nature or 'nature-deficit disorder' is a critical factor in how we behave and respond to issues of environmental impact and sustainability.

It is evident from this study that exemplar sustainable buildings do encourage sustainable behaviour on many levels but buildings can also serve to discourage sustainable behaviour where unintended consequences arise out of poor communication or design decisions. It is clear that people have strong emotional and psychological connections to the buildings they create and inhabit. This is enhanced as the level of involvement in their design and operation increases and can lead to a fundamental change in attitudes, perceptions and ultimately sustainable behaviours (eco-psychological transformation) as awareness of economic, environmental and social sustainability develops.

7.6 Critical Assessment

With sustainable architecture, it is necessary to think differently about both the form and purpose of our buildings and the processes by which they are built. Change in thinking requires that we challenge social, psychological and physiological routines that we have developed and that have seemingly worked well in the past within narrower environmental parameters. Such change is not easy and will be slow and likely to invite some resistance.

This thesis develops an understanding of the dynamic and complex systems and processes by which environmentally responsible action emerges from the interaction of people and the built environment. Buildings and landscapes have a strong influence on our perception of the environment around us. Decisions that result in a poorly designed and operated building can divorce its occupants from a sense of place, its effect on local, regional, national and global ecology, its use of scarce and finite resources with related consequences on climate change and pollution and can have a detrimental effect on the health, well-being and levels of performance of its occupants.

Findings show that the UK building community specialises too much and that it should move more towards an interdisciplinary approach. It was noted that specialised contractors such as PV installers and electricians were so focused on their job, they often did not absorb the messages of sustainability in the project. It was cited that some of the macho behaviour and negative peer pressure on-site limited engagement with environmental issues which ultimately affected the build quality.

The false lessons that unsustainable buildings can impart to us are that resources are infinite, wastefulness is the norm and it is acceptable for us to be removed from natural processes without social, environmental or economic consequences. A sustainable building should be diametrically opposed to this and should explicitly and implicitly enable us to lead more sustainable lifestyles by making sustainable choices the default option. If we are to live more sustainable lifestyles our built environment should be responsive to our needs in a sustainable way while our behaviours should not undermine the potential for our buildings to achieve their sustainable design aspirations.

It is also shown that existing interventions such as building regulations and environmental rating systems do drive sustainable behaviour throughout the construction process, in adopting new and sustainable methods, materials and technologies.

There needs to be a greater understanding of human behaviour change through behavioural science studies specifically, but not exclusively, among designers and building managers who are identified as key stakeholders with the greatest potential to effect behaviour change. This should be incorporated into formal education and working practices.

7.7 Key Recommendations

See Table 7.1.

Table 7.1 Key recommendations

General recommendations
• The decision to build demands early identification of project objectives including function, economic viability, social acceptability and environmental sustainability and the setting of clear behavioural goals
• Use decision-making techniques that can allow both 'hard' and 'soft' issues to be assessed; this should embed sustainability issues at the heart of construction decision-making from the earliest stage in a project
• Companies are coming under increasing pressure to demonstrate social responsibility and this adds a whole new range of issues onto the decision-making agenda
• A new role of carbon/sustainability manager/champion with real power to take responsibility for the carbon performance during the design, construction and operation of the building
• A single organisation with whole life responsibility for a building, ensuring a low carbon approach to design, construction, fit-out, maintenance, refurbishment, retrofitting and even demolition can bring real knowledge to projects enhancing value, education and skills in delivering value, carbon reduction and sustainability in the built environment
• Adopt a whole-life value approach. Build an audit trail into the process to review decisions on whole-life costs, in particular for sustainable decisions. Allocate sufficient time and resources
• Any method of assessing environmental or sustainability performance must be devised and bought into by all the relevant stakeholders
• Wider green economy benefits. The green economy represents an area of substantial potential growth for the built environment professions. Creating a low carbon construction industry would develop skills and expertise that would be of great value to other sectors
• Review and recommend the best methods and lifecycle financial analysis for building projects
• Governments and industry need to work together to identify best practices that stimulate the market for low carbon and energy efficiency measures
• Provide financial incentives and direct funding to lower project costs, e.g. tax credits/ exemptions, remove subsidies for environmentally damaging industries, tax environmentally harmful products, provide low-interest loans for sustainable building projects
• Expand and fund professional training, continuing and formal education opportunities for architects, contractors, realtors, appraisers, developers and lenders
• Provide additional support for sustainable building research and development
• Potential clients need clear explanations of the social, economic and environmental benefits of sustainable and low carbon measures, materials, technologies and methods
• Recognise that different clients are motivated by different sustainability-related issues and the relative importance of different drivers will be unique to each organisation
• Counter misperceptions or exaggeration of the increased cost of sustainable practices
• A key part of developing a procurement strategy directed towards delivering sustainable construction must involve the identification of sustainability-linked risks and the assessment of the most appropriate party to manage these risks
• Delivering sustainable construction is dependent upon adopting the right procurement strategy
• Common language for sustainability is required throughout the construction professions
• Consult openly and willingly with local authorities and the local community at planning stage of construction projects as well as during site operations. Be prepared to change
Design phase recommendations

(continued)

Table 7.1 (continued)

• Early and comprehensive stakeholder involvement (BREEAM assessor, contractors, facilities managers, users and educationalists)
• Late appointment of a sustainability adviser, after fundamental decisions have been taken, can result in a range of design issues that are difficult to overcome
• Consider commitments to achieve issues at BREEAM design stage that will need to be demonstrated at BREEAM post-construction stage
• Alter building contracts to put the responsibility for building performance directly in the hands of both the architect and the service engineer
• Industry should agree a standard method of measuring embodied carbon for use as a design tool, and for the purposes of scheme appraisal
• Interdisciplinary and integrative working practices are needed
• Embed sustainable materials and technologies into buildings and make them 'transparent' to enable users to better understand sustainable principles and the functionality of their buildings
• Shallow-plan building forms, demanding less technically complex and less management-intensive systems
• Embed sustainable and educational aims and objectives into building design, construction, operation and use to encourage greater learning and sustainable behaviour
• Balance technical solutions with human interaction and behaviours
• Behaviour change initiatives should be based on a package of measures. Address both physical/infrastructural ('external') barriers and attitudinal/psychological ('internal') factors
• Participatory design. Encourage participation of local community and building user groups early in the design process
Construction phase recommendations
• Comprehensive training to move sustainable systems, methods, techniques and skills into mainstream construction. Provide induction and training about environmental issues to all site personnel
• Early contractor involvement will ensure that the design does not compromise sustainable construction
• In most construction companies the sustainability manager has a facilitation role, rather than power to drive real change. Sustainable/carbon management needs to be a high level role, on the same basis as the financial manager or the project manager
• A construction-specific accreditation scheme for companies committed to improving their environmental credentials
• Accredited courses for specialist low carbon technologies and techniques to ensure adequate skills, and expertise to avoid undermining the credibility of sustainable building projects by unskilled workers and poor performing materials and technologies
• Include environmental issues in the selection of contactors and require a sustainability/environmental management plan from the main contractor
• Include in specifications requests for contactors' environmental policies, for evidence of environmental innovation on previous projects, and for innovative environmental ideas on the current project
• Ensure environmental issues pervade the site management and practices rather than become side-lined with a few individuals in separate meetings
• Set appropriate targets for environmental performance, including energy, water and waste management

(continued)

Table 7.1 (continued)

• Measure environmental and sustainability performance in order to manage it. Adopt appropriate sustainability indicators, set challenging but achievable targets and benchmark performance against industry best practice
• Report on the material, energy and waste flows associated with products and services over their entire lifecycle and identify changes to operations, which can lead to clear environmental benefit and cost savings
Post-occupancy phase recommendations
• A well-managed project handover (soft landing) with training for occupants and facilities managers on new low carbon systems, materials and technologies, e.g. BEMS. A building should not be considered as complete until it performs in accordance with its design criteria
• Incorporate and embed features that enable the sustainable operation of the building and opportunities for user engagement, e.g. rapid response environments in terms of personal control and complaints monitoring and feedback system
• A need for up-skilling energy, carbon and sustainability knowledge with the maintenance of buildings. The role of sustainable building/facilities manager could have effective power to take responsibility for the energy and carbon operational performance
• A need for raising awareness and training in the operation and use of buildings for occupants in order to maximise energy efficiency, carbon saving and use of sustainability features of buildings
• Do not overestimate the need for future adaptability
• Look upon a building that must be demolished as a source of materials and resources, either on the given project or other projects in the vicinity, or for sale
• Monitor buildings in use to ensure they perform as intended. Use data collected to feed back to operators and use to inform design of new buildings
• Provide documentation for the occupier/user, e.g. operation and maintenance manuals
• Create a document that describes the learning potential of the building that translates environmental prompts, cues or features into learning experiences
• Connect lifestyles and working practices to daily/seasonal cycles
• Publicise sustainable approaches being taken to reduce operational energy use and provide user-friendly performance feedback mechanisms
• Poorly built environments reinforce unsustainable behaviours

Source Author

References

1. Orr DW (2004) The nature of design: ecology, culture and human intention. Oxford University Press, Oxford
2. Stern N (2006) Stern review on the economics of climate change: executive summary. http://www.hm-treasury.gov.uk/sternreview_index.htm. Accessed 28 Jan 2012
3. BRE (2003) Carbon dioxide emissions from non-domestic buildings 2000 and beyond. BRE Press, Watford
4. Frankiwiez J (2009) A low carbon future: an action plan for UK construction. Building 4 Change. www.building4change.com/page.jsp?id=205. Accessed 17 Dec 2009
5. Climate Change Act (2008) Carbon targeting and budgeting. Chapter 27, Part 1. HMSO, London
6. Baird G (2010) Sustainable buildings in practice: what the users think. Routledge
7. Bennetts R, Bordass W (2007) Keep it simple and do it well. Sustain Suppl Build Mag 28:8–11
8. Clune S (2010) Design and behavioural change. The new school. The journal of design strategies. Parsons, New York, pp 68–75
9. Egan J (2004) The Egan review: skills for sustainable communities. Office of the Deputy Prime Minister. HMSO, London
10. Hadi M, Halfhide C (2009) The move to low-carbon design: are designers taking the needs of building users into account? A guide for building designers, operators and users. BRE Press, Watford
11. Hoffman J, Henn R (2008) Overcoming the social and psychological barriers to green building. Organ Environ 21(4):390–419
12. Janda B (2011) Buildings don't use energy: people do. Archit Sci Rev 54:15–22
13. Leaman A (2009) Human factors: the bottom line. Ecolibrium J Austr Instit Refrig Air Condit Heat 2009:36–38
14. Orr DW (2008) Design on the edge. MIT Press, Cambridge, MA
15. Roaf S (2010) Oh, behave, if you want real low carbon buildings. Green Build Mag Autumn 20(2):7
16. Yates A (2003) Sustainable buildings: benefits for occupiers. In: BRE centre for sustainable construction. Information paper 13/03 part 1. BRE Press, Watford
17. Gordon A (1972) Simple societies and complex technologies: designing for survival. In: RIBA annual conference
18. Landman M (1999) Breaking through the barriers to sustainable building: insights from building professionals on government initiatives to promote environmentally sound practices. MSc thesis. TUFTS University, US
19. EESC (2011) Let's speak sustainable construction: a multilingual glossary. European Economic and Social Committee, Brussels

J. L. Clarke, *Sustainable Buildings for Sustainable Behaviour: Theory and Practice*, Green Energy and Technology, https://doi.org/10.1007/978-3-031-98905-6

20. Williamson T, Radford A, Bennetts H (2003) Understanding sustainable architecture. Spon Press, Oxford
21. Steele J (2005) Ecological architecture: a critical history. Thames and Hudson, London
22. Liddell H (2008) Eco-minimalism: the antidote to eco-bling. RIBA Publishing, London
23. Pepper D (1984) The roots of modern environmentalism. Routledge, London
24. Guy S, Farmer G (2000) Contested constructions: the competing logics of green buildings and ethics. In: Fox W (ed) Ethics and the built environment. Routledge, London
25. Pearson D (1991) Making sense of architecture. Archit Rev 1136:68–70
26. Olgyay V (1973) Design with climate: bioclimatic approach to architectural regionalism. Princeton University Press, Princeton
27. Cook SJ, Golton BL (1994) Sustainable development concepts and practice in the built environment: a UK perspective. Sustain Constr CIB TG 16:677–685
28. Kant I (2007) Critique of judgement. Cosimo Classics, New York
29. Healy P (2003) Beauty and the sublime. Sun Publishers, Amsterdam
30. Lee S (2011) Aesthetics of sustainable architecture. 010 Publishers, Rotterdam
31. Kollewe J, Hawkes A (2011) Gherkin architect declares end of London skyscraper boom. http://www.guardian.co.uk/business/2011/apr/20/gherkin-architect-london-skyscraper/print. Accessed 14 July 2011
32. Pastemack A (2009) How green buildings should look: Ken Yeang. http://www.treehugger.com/sustainable-product-design/how-green-buildings-should-look-ken-yeang.html. Accessed 6 Sept 2012
33. Hill G (2011) The aesthetics of architectural consumption. In: Lee S (ed) Aesthetics of sustainable architecture, pp 26–40
34. Commission B (1987) Our common future. Oxford University Press, Oxford
35. Vidal J (2012) Arctic ice shrinks 18% against record, sounding climate change alarm bells. The Guardian Newspaper. http://www.guardian.co.uk/environment/2012/sep/19/arctic-ice-shrinks?CMP=EMCENVEML1631. Accessed 24 Sept
36. National Snow and Ice Data Center (2012) Climate change in the arctic. http://www.nsidc.org/cryosphere/arctic-meteorology/climate_change.html. Accessed 24 Sept 2012
37. United Nations (2011) World population prospects: the 2010 revision. Highlights and Advance Tables. http://esa.un.org/unpd/wpp/Documentation/pdf/WPP2010_Highlights.pdf. Accessed 1 Sept 2012
38. Dunlap RE, Van Liere KD (1978) The new environmental paradigm: a proposed measuring instrument and preliminary results. J Environ Educ 9:10–19
39. Dunlap RE, Van Liere KD (1984) Commitment to the dominant social paradigm and concern for environmental quality. Soc Sci Quart 65:1013–1028
40. Pelham BW (2009) Awareness, opinions about global warming vary worldwide. Gallup World. http://www.gallup.com/poll/117772/Awareness-Opinions-Global-Warming-Vary-Worldwide.aspx#1. Accessed 12 Oct 2012
41. Worldwatch Institute (2013) Vehicle production rises, but few cars are "green". http://www.worldwatch.org/node/5461. Accessed 7 Aug 2013
42. Cohen S (2001) States of Denial: knowing about atrocities and suffering. Polity Press, Cambridge
43. Norgaard KM (2007) Understanding the climate ostrich. http://news.bbc.co.uk/1/hi/sci/tech/7081882.stm. Accessed 10 May 2010
44. Marshall G (2001) The psychology of denial: our failure to act against climate change. http://ecoglobe.ch/motivation/e/clim2922.htm. Accessed 5 March 2012
45. Brown A (2012) If we are to cope with climate change we need a new moral order. http://www.theguardian.com/commentisfree/andrewbrown/2012/aug/21/confront-climate-change-moral. Accessed 24 Aug 2012
46. Clark D (2012) Has the Kyoto protocol made any difference to carbon emissions? http://www.theguardian.com/environment/blog/2012/nov/26/kyoto-protocol-carbon-emissions. Accesses 7 Aug 2013

47. WRAP Waste and Resources Action Programme (2007) Halving construction waste to landfill by 2012. http://www.wrap.org.uk/downloads/Halving_waste_to_lanfill_briefing_note.92f8baa8.4576.pdf. Accessed 19th June 2009
48. Horvath A (2004) Construction materials and the environment. Annu Rev Environ Resour 29:181–204
49. Dixon T (2010) Putting the 'S-word' back into sustainability: can we be more social? In: Oxford institute for sustainable development (OISD). Oxford Brookes University, Oxford
50. DCLG (2007) Building a greener future: policy statement. Department for Communities and Local Government, London
51. CCRA (2012). UK climate change risk assessment. Department for Environment, Food and Rural Affairs. HMSO, London
52. Grondzik WT, Kwok AG, Stein B, Reynolds JS (2010) Mechanical and electrical equipment for buildings, 11th edn. Wiley, New Jersey
53. Lazarus N (2002) Construction materials report: toolkit for carbon neutral developments—part 1. Bioregional Publishing, London
54. Roaf S (2004) Closing the loop: benchmarks for sustainable buildings. RIBA Publishing, London
55. Gonzalez JM, Navarro JG (2006) Assessment of the decrease of CO_2 emissions in the construction field through the selection of materials: a practical case study of three houses of low environmental impact. J Build Environ 41(7):902–909
56. Woodhouse C (2012) London facing killer summers in peril. Evening Standard Newspaper, p 22
57. Edwards B (2012) Mitigation: the built environment and climate change. In: RIBA sustainability hub. http://www.architecture.com/SustainabilityHub/Designstrategies/Introduction/1-0-3-Mitigationthebuiltenvironmentandclimatechange.aspx. Accessed 5 Jan 2012
58. Dripps RD (1999) The first house: myth, paradigm and the task of architecture. MIT Press, Cambridge
59. Perez-Gomez A (1983) Architecture and the crisis of modern science. MIT Press, Cambridge
60. BBC (2011) Brazil Amazon deforestation 'at lowest level in years'. http://www.bbc.co.uk/news/world-latin-america-16048503. Accessed 5 Sept 2012
61. Gray L (2010) Pollution creating acid oceans. The Telegraph. http://www.telegraph.co.uk/earth/earthnews/7223827/Pollution-creating-acid-oceans.html. Accessed 5 Sept 2012
62. Natural England (2011) From first life to the industrial revolution. http://publications.naturalengland.org.uk/file/79018. Accessed 5 Sept 2012
63. Wingspread Conference (1998) Precautionary principle. http://www.sehn.org/ppfaqs.html. Accessed 20 Aug 2012
64. Hajer M (1995) The politics of environmental discourse: ecological modernization and the policy process. Oxford University Press, Oxford
65. Gardiner SM (2011) A perfect moral storm: the ethical tragedy of climate change. Oxford University Press, Oxford
66. Rogers S, Harvey F (2012) Global carbon emissions rise is far bigger than previous estimates. http://www.guardian.co.uk/environment/2012/jun/21/global-carbon-emissions-record. Accessed 5 Sept 2012
67. Spector T (2001) The ethical architect. Princeton Architectural Press, Princeton
68. Bradley A (2007) Don't get homesick. Green Building Press, London
69. Roaf S, Nicol F, Humphreys M, Tuohy D, Boersba S (2010) Twentieth century standards for thermal comfort: promoting high energy buildings. Architect Sci Rev 53:65–77
70. NEF (2010) Good foundations: a low carbon high well-being built environment. New Economics Foundation, London
71. Ellis EV (2010) Building in the 21st century. In: Lecture on building the green economy: university research in sustainable systems. National Building Museum, Washington
72. Nicol J, Roaf S (2007) Adaptive thermal comfort and passive buildings. In: Santamouris M (ed) Passive cooling. James and James Science Publishers, London

73. Humphreys M, Nicol F (1998) Introduction to adaptive comfort. ASHRAE Tech Data Bull 14(1):998
74. Wilson R (1998) Sunlight for human as well as environmental health. In: Lecture to MSc renewable energy and architecture. University of Nottingham
75. Ott JN (2000) The effects of natural and artificial light on man and other living things. Pocket Books, New York
76. Heschong-Mahone (1999) Daylighting in Schools: an investigation into the relationship between daylighting and human performance. Fair Oaks California: Heschong-Mahone Group
77. CIBSE (2002) Code for lighting. In: Chartered institute of building services engineers. Elsevier Science, Oxford
78. Roaf S, Crichton D, Nicol F (2009) Adapting buildings and cities for climate change, 2nd edn. Oxford Architectural Press, Oxford
79. Tuohy P (2008) Air conditioning: the impact of UK regulations, the risks of unnecessary air conditioning and a capability index for non-air conditioned naturally ventilated buildings. In: Proceedings of the 2008 Windsor conference on air conditioning and the low carbon cooling challenge. http://www.nceub.org.uk/index.php?pagename=ResearchWindsorConference2008. Accessed 15 Oct 2011
80. Seppanen O, Fisk WJ (2002) Association of ventilation system type with SBS symptoms in office workers. Indoor Air 12:98–112
81. Smedje G, Norback D (2000) New ventilation systems at select schools in Sweden: effects of asthma and exposure. Arch Environ Health 55(1):18–25
82. EPA (2003) Indoor air quality and student performance. Environmental Protection Agency, Washington DC
83. Buckley J, Schneider M, Shang Y (2011) Los Angeles unified school district school facilities and academic performance. http://www.edfacilities.org/pubs/LAUSD%20Report.pdf. Accessed 25 Nov 2011
84. Coley DA, Greeves R (2004) The effect of low ventilation rates on the cognitive function of a primary school class. In: Centre for energy and the environment. University of Exeter. http://www.architecture.com. Accessed 15 Nov 2011
85. Kukadia V, Ajiboye P, White M (2005) Ventilation and indoor air quality in schools. In: BRE information paper IP 6/05. IHS BRE Press, Bracknell
86. BRE (2009) Green guide to specification. In: An environmental profiling system for building materials and components, 4th edn. Wiley, London
87. Greenspec (2010) Materials guides. http://www.greenspec.co.uk/materials-guides.php. Accessed 17 March 2011
88. Lioy PJ (2011) Dust: the inside story of its role in the September 11th aftermath. Rowman and Littlefield Publishers, New York
89. BRE (2012) Measuring the wellbeing benefits of interior materials. In: BRE information paper IP20/12. IHS BRE Press
90. Steg L, Gifford R (2008) Social psychology and environmental problems. In: Steg L, Buunk AP, Rothengatter T (eds) Applied social psychology: understanding and managing social problems. Cambridge University Press, New York, pp 184–205
91. Shield BM, Dockrell JE (2003) The effects of noise on children at school: a review. J Mech Eng Acoust 10(2):97–116
92. Department of Health (2012) Direct costs of obesity estimated to be £5.1 billion per year. http://www.dh.gov.uk/health/category/policy-areas/public-health/obesity-healthy-living/. Accessed 24 Sept 2012
93. NHS (2012) Half of UK Obese by 2030. http://www.nhs.uk/news/2011/08August/Pages/half-of-uk-predicted-to-be-obese-by-2030.aspx. Accessed 24 Sept 2012
94. IRP Report (2011) Decoupling natural resource use and environmental impacts of economic growth. http://www.unep.org/resourcepanel/. Accessed 12 July 2012
95. Steffen W, Sanderson A, Tyson PD, Jäger J, Matson P, Moore B, Oldfield F, Richardson K, Schellnhuber H-J, Turner BL, Wasson RJ (2004) Global change and the earth system: a planet under pressure. Springer, New York

96. Brown C, Walpole M, Simpson L, Tierney M (2011) Introduction to the UK natural ecosystem assessment. In: The UK national ecosystem assessment technical report. United Nations Environment Programme-World Conservation Monitoring Centre, Cambridge
97. Lea R (2006) Just another excuse for higher taxes. The Daily Telegraph, p 3
98. Muspratt C, Seawright S (2006) Amber alert over green taxes. The Daily Telegraph, p 23
99. Taylor J (2006) Global warming costs and benefits. http://www.cato.org/blog/global-warming-costs-benefits. Accessed 7 Oct 2011
100. Lomborg B (2006) The stern review: the dodgy numbers behind the latest warming scare. Wall Street J 2:25
101. Lawson N (2006) An appeal to reason: a cool look at global warming. Gerald Duckworth and Co., Ltd., London
102. Jackson T (2009) Prosperity without growth: economics for a finite planet. Earthscan, Oxford
103. Victor PA (2008) Managing without growth: slower by design, not disaster (advances in ecological economics). Edward Elgar Publishing Ltd., Cheltenham
104. Melaver M, Mueller P (2009) The green building bottom line: the real cost of sustainable building. Mc Graw-Hill Books
105. World Business Council for Sustainable Development (2007) Policy directions to 2050: a business contribution to the dialogues on cooperative action. Earthprint Ltd
106. Surgenor A, Butterss I (2008) The price of sustainable schools: how much does it cost to achieve BREEAM schools ratings? In: Information paper: IP/08. BRE Press
107. BRE (2005) Costing sustainability: how much does it cost to achieve BREEAM and EcoHomes ratings? In: Information paper: IP4/05. Building research establishment centre for sustainable construction. BRE Press, Watford
108. Quigley J, Eichholtz P, Kok N (2009) Doing well by doing good? An analysis of the financial performance of green office buildings in the USA. Royal Institution of Chartered Surveyors Research, London
109. DCLG (2011) Cost of building to the code for sustainable homes. In: Department for communities and local government. HMSO, London
110. Yates A (2001) Quantifying the business benefits of sustainable buildings. In: Building research establishment. BRE Press, Watford
111. DECC (2012) CRC energy efficiency scheme. http://www.decc.gov.uk/en/content/cms/emissions/crc_efficiency/crc_efficiency.aspx. Accessed 18 Sept 2012
112. Leaman A, Bordass W (1998) PROBE 15: productivity, the killer variables. Build Serv J 1998:41–43
113. Clements-Croome D (2010) Valuing the non-carbon outcomes of design and construction. In: Ecobuild conference seminar. Excel Centre 2–4 March 2010. London
114. Romm JJ (1999) Cool companies: how the best businesses boost profits and productivity by cutting greenhouse gas emissions. Earthscan, London
115. Kats G (2006) Greening America's schools: costs and benefits. United States Green Building Council (USGBC), Washington DC
116. Short A (2010) Glass buildings are set to become pariahs. Build Des 1906:1
117. Dexter AL (1996) Intelligent buildings: fact or fiction? HVAC Res 2(2):105–106
118. Building4Change (2012) Smart walls set to shape sustainable structures. http://www.building4change.com/page.jsp?=1433. Accessed 24 Aug 2012
119. Cole RJ, Brown Z (2009) Human and automated intelligence in comfort provisioning. In: PLEA 2009—26th conference on passive and low energy architecture, Quebec City, Canada, pp 18–21
120. Harper GDJ (2012) Intelligent buildings: sustainable or just lazy? Green Building Magazine 22(2):48–53
121. Owen A (2011) Decisions, decisions: how do real people choose green measures and technology? Green building magazine. Spring 20(4):22–24
122. Jevons WS (1865) The coal question, 2nd edn. Macmillan and Co., London
123. TSB/ESRC (2008) How people use and misuse buildings. In: Technology strategy board/economic and social research council

124. Bell J (2005) Is smart always sustainable in building design and construction. In: Young J, Brandon PS, Sidwell AC (eds) Smart and sustainable built environments. Wiley-Blackwell, New Jersey, pp 176–185
125. Cohen R, Ruyssevelt P, Standeven M, Bordass W, Leaman A (1999) Building intelligence in use: lessons from the Probe project. Usable Buildings Trust
126. DECC (2012) Quantitative research into public awareness, attitudes, and experience of smart meters. In: IPSOS MORI report for the department of energy and climate change
127. Kroner WM (1997) An intelligent and responsive architecture. Autom Constr 6:381–393
128. Lee S-A (2007) Environmental psychology research in practice: thoughts on sustainable architecture and urban design. In: Edgerton E, Romice O, Spencer C (eds) Environmental psychology: putting research into practice. Cambridge Scholars Publishing, Newcastle, pp 12–29
129. Winter D, Koger S (2004) The psychology of environmental problems, 2nd edn. Lawrence Erlbaum Associates, California
130. Stern PC (1997) Toward a working definition of consumption for environmental research and policy. In: Stern PC, Dietz T, Ruttan VR, Socolow RH, Sweeney JL (eds) Environmentally significant consumption: research directions. National Academy Press, Washington DC, pp 12–35
131. Beyond Behaviour Change (2012) http://www.rmit.edu.au/cfd/beyondbehaviour
132. Abroms LC, Maibach E (2008) The effectiveness of mass communication to change public behaviour. Ann Rev Public Health 29:219–234
133. Corner A (2010) Non-environmental behaviour change. Climate Outreach and Information Network (COIN). http://coinet.org.uk/communications-theory-and-tools. Accessed 15 June 2011
134. Das E, De Wit J, Stroebe W (2003) Fear appeals motivate the acceptance of action recommendations: evidence for a positive bias in the processing of persuasive messages. Person Soc Psychol Bull 29:650–664
135. Lorenzoni I, Nicholson-Cole S, Whitmarsh L (2007) Barriers to engaging with climate change among the UK public and their policy implications. Global Environ Change 17:445–459
136. Bhamra T (2009) Design for sustainable living: turning values into actions. In: Lecture to design research group
137. Bhamra T, Lilley D, Tang T (2011) Design for sustainable behaviour: using products to change consumer behaviour. Des J 14(4):427–445
138. Wever R, van Kuijk J, Boks C (2008) User-centred design for sustainable behaviour. Int J Sustain Eng 1(1):9–20
139. Chick A, Micklethwaite P (2011) Design for sustainable change: how design and designers can drive the sustainability agenda. AVA Publishing
140. Brezet H, Van Hemel C (1997) Ecodesign: a promising approach to sustainable production and consumption. UNEP, Paris
141. Stevels A, Griese H (2004) Electronics goes green: current and future issues. In: Reichl H, Griese H, Potter H (eds) Electronics goes green: driving forces for future electronics. IEEE, Berlin, pp 45–54
142. Abele E, Anderl R, Birkhofer H (2005) Environmentally-friendly product development: methods and tools. Springer, London
143. TSB (2010) Integrating smart meters into smart homes. Technology Strategy Board
144. Van Dam SS, Bakker CA, van Hal JDM (2010) Home energy monitors: impact over the medium term. Build Res Inform 38(5):458–469
145. Ottman J (2004) Green marketing: opportunity for innovation. BookSurge, Charleston
146. Kohler W (1970) Gestalt psychology: an introduction to new concepts in modern psychology. Liverlight, New York
147. Lewin K (1951) Formalization and progress in psychology: In: Cartwright D (ed) Field theory in social science. Harper, New York
148. Zeisel J (1984) Inquiry by design: tools for environment-behaviour research. Cambridge University Press, Cambridge

149. Leaman A, Bordass W (2001) Assessing building performance in use 4: the PROBE occupant surveys and their implications. Build Res Inform 29(2):129–143
150. Maiteny P (2005) Education for sustainability and development: psycho-emotional blocks and catalysts. Develop Educ J 11(2):12–14
151. Pichert D, Katsikopoulos KV (2008) Green defaults: information presentation and pro-environmental behavior. J Environ Psychol 28(1):63–73
152. Hardin S (1968) The tragedy of the commons. Science 162(13):1243–1248
153. Aarts H, Verplanken B, Van Knippenberg A (1998) Predicting behaviour from actions in the past: repeated decision making or a matter of habit? J Appl Soc Psychol 28:1355–1374
154. Verplanken B, Wood W (2006) Interventions to break and create consumer habits. J Public Policy Market 25:90–103
155. Ajzen I (1985) From intentions to action: a theory of planned behaviour. In: Kuhl J, Beckham J (eds) Action-control: from cognition to behaviour. Springer, Heidelberg, pp 40–58
156. Verplanken B, Aarts H, Van Knippenberg A, Moonen A (1998) Habit versus planned behaviour: a field experiment. Br J Soc Psychol 37:111–128
157. Becker LJ, Seligman C (1978) Reducing air conditioning waste by signaling it is cool outside. Person Soc Psychol Bull 4:412–415
158. Geller ES, Winnett PA, Everett PB (1982) Preserving the environment: new strategies for behavior change. Pergamon, New York
159. Lockton D (2010) What is design with intent? http://www.danlockton.com. Accessed 16 July 2012
160. Winnett RA, Leckliter IN, Chinn D, Stahl B, Love SQ (1985) Effects of television modeling on residential energy consumption. J Appl Behav Anal 18:33–44
161. Geller ES (2001) The psychology of safety handbook. CRC Press, Boca Raton
162. Geller ES (2002) The participation factor: how to increase involvement in occupational safety. American Society of Safety Engineers, Des Plaines
163. Cialdini RB (2001) Influence: science and practice, 4th edn. Allyn and Bacon, Boston
164. Kohlenberg R, Phillips T (1973) Reinforcement and rate of litter depositing. J Appl Behav Anal 6:391–396
165. Van Houwelingen JH, Van Raaij FW (1989) The effect of goal-setting and daily electronic feedback on in-home energy use. J Consum Res 16:98–105
166. Newholme N (1990) Implications of attitude and behaviour research for environmental conservation. J Environ Educ 22(1):26–32
167. Olsen ME (1981) Consumers' attitudes toward energy consumption. J Soc 37:108–131
168. Dwyer WO, Leeming FC, Cobern MK, Porter BE, Jackson JM (1993) Critical review of behavioural interventions to preserve the environment: research since 1980. Environ Behav 25:275–321
169. Heberlein TA (1975) Conservation information: the energy crisis and electricity consumption in an apartment complex. Energy Syst Policy 1:105–117
170. Kempton W, Harris CK, Keith JC, Weil JS (1985) Do consumers know what works in energy conservation? Marriage Fam Rev 9:115–133
171. Thaler RH, Sunstein CR (2008) Nudge: improving decisions about health, wealth and happiness. Yale University Press, New York
172. Kolbert E (2008) Buckminster fuller. The New Yorker, p 11
173. Slovic P (1978) Perception of risk. Science 236:280–285
174. Wade-Benzoni K (1996) Intergenerational justice. Northwestern University, Evanston, IL
175. Margai FM (1997) Analyzing changes in waste reduction behavior in a low-income urban community following a public outreach program. J Environ Behav 29(6):769–792
176. BBC (2009) Bang goes the theory: the human power station. Broadcast on 3(12):2009
177. Monbiot G (2008) Our craving for deception. The Guardian Newspaper, p 29
178. Merritt AC, Effron DA, Monin B (2010) Moral self-licensing: when good frees us to be bad. Soc Psychol Compass 4(5):344–357
179. Jenks M, Dempsey N (eds) (2005) Future forms and design for sustainable cities. Architectural Press, Oxford

180. Shove E (2009) Climate change and everyday life. In: Britain in 2010 annual magazine of the economic and social research council, p 20
181. Wall K, Dockrell J, Peacey N (2008) Primary schools: the built environment research survey. University of Cambridge, Cambridge
182. Rutter M, Maughan B, Mortimore P, Ouston J (1979) Fifteen thousand hours: secondary schools and their effect on children. Open Books, London
183. Blair T (2004) PM speech on climate change. Archive No. 10 downing street London. http://www.pm.gov/output/page 6333.asp. Accessed 20 Sept 2004
184. BBC (2009) Costing the earth: energy high. British Broadcasting Corporation Radio 4 Broadcast
185. Sterling S (2002) Sustainable education: revisioning learning and change. Green Books, Totnes
186. Simpson V (2008) This will be your school. In: The Guardian newspaper, p 5
187. Wright E (2011) Freedom fight: toby young interview. Build Mag. http://m.building.co.uk/sectors/education/freedom-fighter-toby-young-interview/5005207.article. Accessed 4 Sept 2011
188. Building4change (2012) Government slashes costs and space with standardised school design. http://www.building4change.com/pagejsp?id=1495. Accessed 5 Oct 2012
189. Jodidio P (2006) Architecture nature. Restel, London
190. Sassi P (2006) Strategies for sustainable architecture. Taylor and Francis, Oxford
191. Taylor A (2009) Linking architecture and education: sustainable design and learning environments. University of Mexico Press, New Mexico
192. Blenkin GA, Edwards G, Kelly AV (1992) Recent and emergent theoretical perspectives. Change Curricul 12:39–69
193. Firth R (2008) Environmental education/education for sustainable development: making a difference? School of education presentation. Nottingham University
194. Bonnett M (2007) Environmental education and the issue of nature. J Curricul Stud 39(6):707–721
195. Firth R, Winter C (2007) Constructing education for sustainable development: the secondary school geography curriculum and initial teacher training. Environ Educ Res 13(5):599–619
196. NTL (1954) The learning pyramid. institute for applied behavioural science. National Training Laboratory, Maine
197. Fishbein M, Ajzen I (1980) Understanding attitudes and predicting social behaviour. Prentice-Hall, Englewood Cliffs
198. Kobet R (2003) Empowering learning through natural, human, and building ecologies. http://www.designshare.com/Research/Kobet/learning_ecology_2.htm. Accessed 4 June 2012
199. Davis M, Hawley D, McMullan B, Spilka G (1997) Design as a catalyst for learning. Association for Supervision and Curriculum Development, Alexandria
200. Pallack MS, Cook DA, Sullivan JJ (1980) Commitment and energy conservation. In: Bickman L (ed) Applied social psychology annual, pp 235–253
201. O'Rourke Swift D (200) Effects of students population density on academic achievements in Georgia elementary schools. PhD dissertation. http://www.eric.ed.gov
202. Tiburcio T, Finch EF (2005) The impact of an intelligent classroom on pupils' interactive behaviour. Facilities 23(6):262–278
203. Caine RN, Caine G (1991) Making connections: teaching and the human brain. Association for Supervision and Curriculum Development, Alexandria
204. Locker FM, Olsen S (2004) Flexible school facilities. http://www.designshare.com/Research/Locker/Flexible.Schools.asp. Accessed 1 Oct 2011
205. Blatchford P, Bassett P, Goldstein H (2003) The class size debate: is smaller better? Oxford University Press, Oxford
206. CABE (2010) Does school design matter? http://webarchive.nationalarchives.gov.uk/20110118095356/http:www.cabe.org.uk/news/does-school-design-matter. Accessed 18 Oct 2010
207. Illinois Healthy Schools Campaign (2003) Apparently size doesn't matter: two Illinois school districts show successful IAQ management. School Health Watch. http://healthyschoolscampaign.org/news/newsletter/2003-summer_HSC-newsletter.pdf. Accessed 28 Sept 2012

208. Schneider M (2002) Public school facilities and teaching: Washington DC and Chicago. A report prepared for the neighbourhood capital budget group (NCBG). http://www.ncbg.org/press/press111302.htm. Accessed 5 Oct 2012
209. Paladino and Company (2005) Washington high performance school buildings: report to legislature. Paladino and Company, Washington
210. Doll W (2005) Green design experiences: a case study. PhD dissertation for the University of Pennsylvania 2005
211. OFSTED (2009) Education for sustainable development. In: Improving schools: improving lives. Manchester: office for standards in education, children's services and skills. HMSO, London
212. Birney A, Reed J (2009) Sustainability and renewal: findings from the leading sustainable schools research project. National College for Leadership of Schools and Children's Services, Nottingham
213. Thomas G, Thomson G (2004) A child's place: why environment matters to children. Green Alliance/DEMOS, London
214. Hicks D, Holden C (2007) Remembering the future: what do children think? Environ Educ Res 13(4):501–512
215. Alexander R (ed) (2009) Children, their world, their education. Routledge, London
216. Gayford C (2009) Learning for sustainability: from the pupils' perspective. World Wide Fund for Nature, Godalming. http://assets.wwf.org.uk/downloads/wwf_report_final_web.pdf. Accessed 2 Oct 2010
217. Murphy C, Thorne A (2010) Health and productivity benefits of sustainable schools: a review. Building Research Establishment Press, Watford
218. Building4Change (2012) Classrooms' influence on pupil progress could be 25 per cent. http://www.building4change.com/pagejsp?id=1561. Accessed 21 Nov 2012
219. Ryghaug M (2003) Towards a sustainable aesthetics. In: Architects constructing energy efficient buildings. PhD thesis. Norwegien University of Science and Technology
220. Fallan K (2008) Architecture in action: travelling with actor-network theory in the land of architectural research. Architect Theory Rev 31:80–96
221. Hayles C, Holdsworth S (2008) Curriculum change for sustainability. J Educ Built Environ 3(1):25–48
222. Egri C, Pinfield L (1996) Handbook of organization studies. Sage, London, pp 459–483
223. Walker P (2001) From stategic adviser to design subcontractor and back again. In: Abley I, Heartfield J (eds) Sustaining architecture in the anti-machine age. Wiley, London
224. Frederick M (2007) 101 things i learned in architecture school. MIT Press, Cambridge
225. Cowan R (2010) Plandemonium. The Urban Design Group, London
226. Hunting SA, Tilbury D (2006) Shifting towards sustainability: six insights into successful organisational change for sustainability. In: Australian research institute in education for sustainability (ARIES). Government department of the environment and heritage. ARIES, Sydney
227. CIRIA (2001) Sustainable construction procurement: a guide to delivering environmentally responsible projects—C571. Construction Industry Research and Information Association, London
228. Carbon Smart (2009) Human resources: a critical piece in the green office puzzle. http://www.carbonsmart.co.uk/index.php?q=human-resources-green-office. Accessed 14th Dec 2009
229. Allison G (1971) Essence of decision. Harper Collins, Boston
230. Orr DW (2008) Our great work. Resurgence. No. 248:29–222
231. Kempton W, Boster JS, Hartley JA (1995) Environmental values in American culture. MIT Press, Cambridge
232. Kibert CJ (2007) Sustainable construction: green building design and delivery. Wiley, Hoboken
233. Mintzberg H (1979) The structuring of organisations. Prentice Hall, Englewood Cliffs
234. Chamberlain L (2008) Serving architects, consultants in everything green become mainstays. The New York Times, New York. http//www.nytimes.com/2008/08/27/business/27green.html?scp=1&sq=sustainability%20consultant&st=cse. Accessed 1 Sept 2008

235. Howard-Grenville J, Hoffman A (2003) The importance of cultural framing to the success of social initiatives in business. Acad Manag Execut 17(2):70–74
236. Scott WR, Meyer J (1992) The organization of societal sectors. In: Meyer JW, Scott WR (eds) Organizational environments: ritual and rationality. Sage, Thousand Oaks, pp 129–154
237. Easterbrook G (1995) A moment on the earth. Viking, New York
238. Tenbrunsel AE et al (1997) The dysfunctional aspects of environmental standards. Environ Ethics Behav Psychol Environ Valuat Degrad 15:105–121
239. Lovins A, Lovins H (1997) Climate: making sense and making money. Rocky Mountain Institute, Snowmass
240. Zucker L (1983) Organizations as institutions. In: Bacharach S (ed) Research in the sociology of organizations. JAI Press, Greenwich, pp 1–47
241. Moon S, De Leon P (2007) Contexts and corporate voluntary and environmental behaviours: examining the EPA's green lights voluntary program. Organ Environ 20:480–496
242. Berkhout T, Rowlands IH (2007) The voluntary adoption of green electricity by Ontario-based companies: the importance of organizational values and organizational context. Organ Environ 20(3):281–303
243. Natural Marketing Institute (2007) 2006 edition of understanding LOHAS market report: a focus on green building. NMI Publishing, Harleysville
244. Latham M (1994) Constructing the team: final report of the government/industry review of procurement and contractual arrangements in the UK construction industry. Department of the Environment. HMSO, London
245. Office of Government Commerce (2007). The integrated project team: teamworking and partnering. Achieving excellence in construction procurement guide 5. OGC, London
246. Oliver M (2012) Why housing needs a lifespan requirement. http://building4change.com/page.jsp?id=1497. Accessed 20th Oct 2012
247. Public Sector Sustainability (2012) New report proves value of BREEAM beyond capital outlay. Abbey Publishing and Exhibitions Ltd., London
248. Jansz A, Mistry V, Yates A (2005) Putting a price on sustainability. In: BRE trust report FB10. IHS BRE Press, Bracknell
249. McKenzie-Mohr D, Smith W (1999) Fostering sustainable behaviour: an introduction to community-based social marketing. New Society Publications, British Columbia
250. Walker JA (1989) Design history and the history of design. Pluto, London, p 183
251. Addis B, Talbot R (2001) Sustainable construction procurement: a guide to delivering environmentally responsible projects—C571. CIRIA, London
252. Reed WG, Boecker J, Horst S, Keiter AL, Sheffer M, Toevs B (2009) The integrative design guide to green building: redefining the practice of sustainability. Wiley, New York
253. Flood A (2011) JCT consults on life cycle issues. http://www.building4change.com/page.jsp?id=1067. Accessed10 Oct 2011
254. Gieryn T (2002) What buildings do. Theory Soc 31(1):35–74
255. Latour B (2005) Reassembling the social: an introduction to actor-network theory. Oxford University Press, Oxford
256. Colomina B (1999) Collaborations: the private life of modern architecture. J Soc Archit Hist 58(3):462–463
257. Fox W (2000) Towards an ethics (or at least a value theory). In: Fox W (ed) Ethics and the built environment. Routledge, London
258. Bell PA, Fisher JD (2001) Environmental psychology. W.B Saunders, Philadelphia
259. Bishop P (2011) The Bishop review: the future of design in the built environment. Design Council and CABE
260. Katz D, Kahn RL (1966) The social psychology of organisations. Wiley, New York
261. Bordass W (2009) The building services brief of the future. Facil Perspect J Facil Assoc Austr 35:548
262. Sanoff H (2000) Community participation methods in design and planning. Wiley, New York
263. Wheeler A (2009) Can we design schools to encourage lifestyle change? Participation and sustainable behaviour. Int J Environ Cult Econ Soc Sustain 5(5):1–16

264. Macnaghten P (2001) Sustainable development and everyday life. In: Abley I, Heartfield J (eds) Sustaining architecture in the anti-machine age. Wiley, Chichester, pp 86–93
265. Scrivens R (ed) (2011) Weak green credentials are not a deal breaker for procurement managers. Public sector sustainability. Abbey Publishing and Exhibitions Ltd., London
266. UKGBC (2009) Sustainability training and education report. UK Green Building Council, London
267. Farrar M (2011) Green skills essential to carbon-conscious building industry. Guardian Newspapers Ltd. http://www.guardian.co.uk/public-leaders-network/2011/jul/13/green-skills-building-industry/print. Accessed 18 July 2011
268. Hall J (2012) Green deal 'in tatters' as no-one registers. The Daily Telegraph, p 1
269. Building4Change (2012) Think low carbon centre opens at Barnsley college. http://www.building4change.com/page.jsp?id=1176. Accessed 24 Feb 2012
270. Pears A (2012) Cafe van: sustaining builders. www.cafevan.org. Accessed 31 Oct 2012
271. BSRIA (2009) The soft landings framework for better briefing, design, handover and building performance in use. In: BSRIA paper BG4/2009
272. Kenrick V (2011) The case for the role of energy and sustainability manager. Sustain Facil Manag 2011:14–15
273. Blyth A, Gilby A (2006) Guide to post occupancy evaluation. University of Westminster School of Architecture and The Built Environment, HEFCE
274. Fulcher M (2011) King: architects could be sued for poor performing green buildings. http://www.architectsjournal.co.uk/news/daily-news/king-architects-could-be-sued-for-poor-performing-green-buildings/8616760. Accessed 13 July 2011
275. Carbon Trust (2011) Cutting workplace carbon, not competitiveness. Low Carbon Workplace Ltd

GPSR Compliance

The European Union's (EU) General Product Safety Regulation (GPSR) is a set of rules that requires consumer products to be safe and our obligations to ensure this.

If you have any concerns about our products, you can contact us on ProductSafety@springernature.com

In case Publisher is established outside the EU, the EU authorized representative is:

Springer Nature Customer Service Center GmbH
Europaplatz 3
69115 Heidelberg, Germany

Batch number: 10755466

Printed by Printforce, the Netherlands